Cosmology

Cosmology

Matthew Turner

STATES
ACADEMIC PRESS
www.statesacademicpress.com

Cosmology
Matthew Turner
ISBN: 978-1-63989-129-0 (Hardback)

STATES
ACADEMIC PRESS

Published by States Academic Press,
109 South 5th Street,
Brooklyn, NY 11249, USA

Cataloging-in-Publication Data

Cosmology / Matthew Turner.
p. cm.
Includes bibliographical references and index.
ISBN 978-1-63989-129-0
1. Cosmology. 2. Astronomy. 3. Universe. I. Turner, Matthew.
QB981 .C67 2022
523.1--dc23

For more information regarding States Academic Press and its products, please visit the publisher's website www.statesacademicpress.com

Table of Contents

Preface

This book aims to help a broader range of students by exploring a wide variety of significant topics related to this discipline. It will help students in achieving a higher level of understanding of the subject and excel in their respective fields. This book would not have been possible without the unwavered support of my senior professors who took out the time to provide me feedback and help me with the process. I would also like to thank my family for their patience and support.

The field of astronomy that focuses on the study of origin and evolution of the universe is known as cosmology. It is concerned with the study of the eventual fate of the universe. This field focuses on studying the universe as a whole rather than studying the individual celestial bodies. This discipline can be divided into physical cosmology, philosophical cosmology and religious or mythological cosmology. Physical cosmology deals with the study of large-structures and dynamics of the universe. Philosophical cosmology is concerned with the world as the totality of time, space and other phenomena. Religious or mythological cosmology deals with the study of beliefs based on religious and mythological traditions and cultures. This field draws significantly from other areas such as physics and astrophysics. It is used in other disciplines such as physics, space and time, metaphysics and philosophy. This book is a valuable compilation of topics, ranging from the basic to the most complex theories and principles related to this discipline. While understanding the long-term perspectives of the topics, the book makes an effort in highlighting their impact as a modern tool for the growth of the discipline. As cosmology is emerging at a rapid pace, the contents of this book will help the readers understand the modern concepts and applications of the subject.

A brief overview of the book contents is provided below:

Chapter - The Universe

The universe is the cosmic system of matter and energy. It includes planets, stars, galaxies, and all other forms of matter and energy. Multiverse, early universe, Friedmann universe, etc. are some of the concepts studied under the domain of universe. This chapter has been carefully written to provide an easy understanding of the related concepts of universe.

Chapter - Cosmology: A Comprehensive Study

The branch of astronomy that deals with the study of origin and evolution of universe is termed as cosmology. Dualistic cosmology involves diverse views from different traditional and scriptural religions. The topics elaborated in this chapter will help in gaining a better perspective of cosmology.

Chapter - Physical Cosmology

The branch of cosmology that deals with the study of largest-scale dynamics and structure of the universe is referred to as physical cosmology. Dark matter, plasma cosmology, observational cosmology, Copernican principle, inhomogeneous cosmology, etc. fall under its domain. This chapter discusses physical cosmology in detail.

Chapter - The Big Bang

The Big Bang is a cosmological model that describes the evolution and expansion of universe from a very high density and high-temperature state. Lambda-CDM model, cosmological horizon, timeline of the universe, etc. are a few of its aspects. This chapter sheds light on the subject of the Big Bang for a thorough understanding of the subject.

Chapter - Galaxies

A gravitationally bound system of stars, stellar remnants, interstellar gas, dust, and dark matter is termed as a galaxy. Some of the galaxies are Milky Way, Andromeda galaxy, elliptical galaxy, spiral galaxy, starburst galaxy, etc. This chapter closely examines these galaxies to provide an extensive understanding of the subject.

Matthew Turner

1
The Universe

The universe is the cosmic system of matter and energy. It includes planets, stars, galaxies, and all other forms of matter and energy. Multiverse, early universe, Friedmann universe, etc. are some of the concepts studied under the domain of universe. This chapter has been carefully written to provide an easy understanding of the related concepts of universe.

The Universe is everything we can touch, feel, sense, measure or detect. It includes living things, planets, stars, galaxies, dust clouds, light, and even time. Before the birth of the Universe, time, space and matter did not exist. The Universe contains billions of galaxies, each containing millions or billions of stars. The space between the stars and galaxies is largely empty. However, even places far from stars and planets contain scattered particles of dust or a few hydrogen atoms per cubic centimeter. Space is also filled with radiation (e.g. light and heat), magnetic fields and high energy particles (e.g. cosmic rays).

The Universe is incredibly huge. It would take a modern jet fighter more than a million years to reach the nearest star to the Sun. Travelling at the speed of light (300,000 km per second), it would take 100,000 years to cross our Milky Way galaxy alone.

No one knows the exact size of the Universe, because we cannot see the edge – if there is one. All we do know is that the visible Universe is at least 93 billion light years across. (A light year is the distance light travels in one year – about 9 trillion km).

The Universe has not always been the same size. Scientists believe it began in a Big Bang, which took place nearly 14 billion years ago. Since then, the Universe has been expanding outward at very high speed. So the area of space we now see is billions of times bigger than it was when the Universe was very young. The galaxies are also moving further apart as the space between them expands.

The best-supported theory of our universe's origin centers on an event known as the big bang. This theory was born of the observation that other galaxies are moving away from our own at great speed in all directions, as if they had all been propelled by an ancient explosive force.

A Belgian priest named Georges Lemaître first suggested the big bang theory in the 1920s, when he theorized that the universe began from a single primordial atom. The idea received major boosts from Edwin Hubble's observations that galaxies are speeding away from us in all directions, as well as from the 1960s discovery of cosmic microwave radiation—interpreted as echoes of the big bang—by Arno Penzias and Robert Wilson.

Further work has helped clarify the big bang's tempo. Here's the theory: In the first 10^-43 seconds of its existence, the universe was very compact, less than a million billion billionth the size of a single atom. It's thought that at such an incomprehensibly dense, energetic state, the four fundamental forces—gravity, electromagnetism, and the strong and weak nuclear forces—were forged into a single force, but our current theories haven't yet figured out how a single, unified force would work. To pull this off, we'd need to know how gravity works on the subatomic scale, but we currently don't.

It's also thought that the extremely close quarters allowed the universe's very first particles to mix, mingle, and settle into roughly the same temperature. Then, in an unimaginably small fraction of a second, all that matter and energy expanded outward more or less evenly, with tiny variations provided by fluctuations on the quantum scale. That model of breakneck expansion, called inflation, may explain why the universe has such an even temperature and distribution of matter.

After inflation, the universe continued to expand but at a much slower rate. It's still unclear what exactly powered inflation.

Bright blue Earth looms over a star field that contains the oldest known planet in the Milky Way. The ancient planet is thought to be about 13 billion years old, a mere billion years younger than the estimated age of the universe. Its discovery, made using the Hubble Space Telescope, is evidence that planets began forming relatively soon after the big bang.

Aftermath of Cosmic Inflation

As time passed and matter cooled, more diverse kinds of particles began to form, and they eventually condensed into the stars and galaxies of our present universe.

By the time the universe was a billionth of a second old, the universe had cooled down enough for the four fundamental forces to separate from one another. The universe's fundamental particles also formed. It was still so hot, though, that these particles hadn't yet assembled into many of the subatomic particles we have today, such as the proton. As the universe kept expanding, this piping-hot primordial soup—called the quark-gluon plasma—continued to cool. Some particle colliders, such as CERN's Large Hadron Collider, are powerful enough to re-create the quark-gluon plasma.

Radiation in the early universe was so intense that colliding photons could form pairs of particles made of matter and antimatter, which is like regular matter in every way except with the opposite electrical charge. It's thought that the early universe contained equal amounts of matter and antimatter. But as the universe cooled, photons no longer packed enough punch to make matter-antimatter pairs. So like an extreme game of musical chairs, many particles of matter and antimatter paired off and annihilated one another.

Somehow, some excess matter survived—and it's now the stuff that people, planets, and galaxies are made of. Our existence is a clear sign that the laws of nature treat matter and antimatter slightly differently. Researchers have experimentally observed this rule imbalance, called CP violation, in action. Physicists are still trying to figure out exactly how matter won out in the early universe.

Building Atoms

Within the universe's first second, it was cool enough for the remaining matter to coalesce into protons and neutrons, the familiar particles that make up atoms' nuclei. And after the first three minutes, the protons and neutrons had assembled into hydrogen and helium nuclei. By mass, hydrogen was 75 percent of the early universe's matter, and helium was 25 percent. The abundance of helium is a key prediction of big bang theory, and it's been confirmed by scientific observations.

Despite having atomic nuclei, the young universe was still too hot for electrons to settle in around them to form stable atoms. The universe's matter remained an electrically charged fog that was so dense, light had a hard time bouncing its way through. It would take another 380,000 years or so for the universe to cool down enough for neutral atoms to form—a pivotal moment called recombination. The cooler universe made it transparent for the first time, which let the photons rattling around within it finally zip through unimpeded.

We still see this primordial afterglow today as cosmic microwave background radiation, which is found throughout the universe. The radiation is similar to that used to transmit TV signals via antennae. But it is the oldest radiation known and may hold many secrets about the universe's earliest moments.

From the First Stars to Today

There wasn't a single star in the universe until about 180 million years after the big bang. It took that long for gravity to gather clouds of hydrogen and forge them into stars. Many physicists think that vast clouds of dark matter, a still-unknown material that outweighs visible matter by more than five to one, provided a gravitational scaffold for the first galaxies and stars.

Once the universe's first stars ignited, the light they unleashed packed enough punch to once again strip electrons from neutral atoms of the universe called reionization. In February 2018, an Australian team announced that they may have detected signs of this "cosmic dawn." By 400 million years after the big bang, the first galaxies were born. In the billions of years since, stars, galaxies, and clusters of galaxies have formed and re-formed—eventually yielding our home galaxy, the Milky Way, and our cosmic home, the solar system.

Even now the universe is expanding, and to astronomers' surprise, the pace of expansion is accelerating. It's thought that this acceleration is driven by a force that repels gravity called dark energy. We still don't know what dark energy is, but it's thought that it makes up 68 percent of the universe's total matter and energy. Dark matter makes up another 27 percent. In essence, all the matter you've ever seen—from your first love to the stars overhead—makes up less than five percent of the universe.

Planets

Planet is any relatively large natural body that revolves in an orbit around the Sun or around some other star and that is not radiating energy from internal nuclear fusion reactions. In addition to the above description, some scientists impose additional constraints regarding characteristics such as size (e.g., the object should be more than about 1,000 km [600 miles] across, or a little larger than the largest known asteroid, Ceres), shape (it should be large enough to have been squeezed by its own gravity into a sphere—i.e., roughly 700 km [435 miles] across, depending on its density), or mass (it must have a mass insufficient for its core to have experienced even temporary nuclear fusion). As the term is applied to bodies in Earth's solar system, the International Astronomical Union (IAU), which is charged by the scientific community with classifying astronomical objects, lists eight planets orbiting the Sun; in order of increasing distance, they are Mercury, Venus, Earth, Mars, Jupiter, Saturn, Uranus, and Neptune. Pluto also was listed as a planet until 2006. Until the close of the 20th century, the only planets to be recognized were components of Earth's solar system. At that time astronomers confirmed that other stars have objects that appear to be planets in orbit around them.

The eight planets of the solar system and Pluto, in a montage of images scaled to show the approximate sizes of the bodies relative to one another. Outward from the Sun, which is represented to scale by the yellow segment at the extreme left, are the four rocky terrestrial planets (Mercury, Venus, Earth, and Mars), the four hydrogen-rich giant planets (Jupiter, Saturn, Uranus, and Neptune), and icy, comparatively tiny Pluto.

Planets of other Stars

The planets and other objects that circle the Sun are thought to have formed when part of an interstellar cloud of gas and dust collapsed under its own gravitational attraction and formed a disk-shaped nebula. Further compression of the disk's central region formed the Sun, while the gas and dust left behind in the midplane of the surrounding disk eventually coalesced to form ever-larger objects and, ultimately, the planets. (See solar system: Origin of the solar system.) Astronomers have long wondered if this process of planetary formation could have accompanied the birth of stars other than the Sun. In the glare of their parent stars, however, such small, dim objects would not be easy to detect directly in images made with telescopes from Earth's vicinity. Instead, astronomers concentrated on attempting to observe them indirectly through the gravitational effects they exert on their parent stars. After decades of searching for such extrasolar planets, astronomers in the early 1990s indirectly identified three planets circling a pulsar (i.e., a rapidly spinning neutron star) called PSR B1257+12. The first discovery of a planet revolving around a star more like the Sun came in 1995 with the announcement of the existence of a massive planet orbiting the star 51 Pegasi. In the first 15 years after these initial discoveries, about 200 planets around other stars were known, and in 2005 astronomers obtained the first direct infrared images of what were interpreted to be extrasolar planets. In size these objects range from a fraction of the mass of Jupiter to more than a dozen times its mass. Astronomers have yet to develop a rigorous, generally accepted definition of planet that will successfully accommodate extrasolar planets and distinguish them from bodies that are more starlike in character (e.g., brown dwarfs).

Star

A star is a massive, luminous ball of plasma that is held together by its own gravity. The nearest star to Earth is the Sun, which is the source of most of the energy on Earth. Other stars are visible in the night sky, when they are not outshone by the Sun. For most of its life, a star shines due to thermonuclear fusion in its core releasing energy that traverses the star's interior and then radiates into outer space. Almost all elements heavier than hydrogen and helium were created by fusion processes in stars.

The Pleiades, an open cluster of stars in the constellation of Taurus.

Astronomers can determine the mass, age, chemical composition and many other properties of a star by observing its spectrum, luminosity and motion through space. The total mass of a star is the principal determinant in its evolution and eventual fate. Other characteristics of a star are determined by its evolutionary history, including the diameter, rotation, movement and temperature. A plot of the temperature of many stars against their luminosities, known as a Hertzsprung-Russell diagram (H–R diagram), allows the age and evolutionary state of a star to be determined.

Binary and multi-star systems consist of two or more stars that are gravitationally bound, and generally move around each other in stable orbits. When two such stars have a relatively close orbit, their gravitational interaction can have a significant impact on their evolution. Stars can form part of a much larger gravitationally bound structure, such as a cluster or a galaxy.

Historically, stars have been important to civilizations throughout the world. They have been used in religious practices and for celestial navigation and orientation. Many ancient astronomers believed that stars were permanently affixed to a heavenly sphere, and that they were immutable. By convention, astronomers grouped stars into constellations and used them to track the motions of the planets and the inferred position of the Sun. The motion of the Sun against the background stars (and the horizon) was used to create calendars, which could be used to regulate agricultural practices. The Gregorian calendar, currently used nearly everywhere in the world, is a solar calendar based on the angle of the Earth's rotational axis relative to the nearest star, the Sun.

The oldest accurately dated star chart appeared in Ancient Egypt in 1,534 B.C.E. Islamic astronomers gave to many stars Arabic names which are still used today, and they invented numerous astronomical instruments which could compute the positions of the stars. In the 11th century, Abū Rayhān al-Bīrūnī described the Milky Way galaxy as multitude of fragments having the properties of nebulous stars, and also gave the latitudes of various stars during a lunar eclipse in 1019.

In spite of the apparent immutability of the heavens, Chinese astronomers were aware that new stars could appear. Early European astronomers such as Tycho Brahe identified new stars in the night sky (later termed novae), suggesting that the heavens were not immutable. In 1584 Giordano Bruno suggested that the stars were actually other suns, and may have other planets, possibly even Earth-like, in orbit around them, an idea that had been suggested earlier by such ancient Greek philosophers as Democritus and Epicurus. By the following century the idea of the stars as distant suns was reaching a consensus among astronomers. To explain why these stars exerted no net gravitational pull on the solar system, Isaac Newton suggested that the stars were equally distributed in every direction, an idea prompted by the theologian Richard Bentley.

The Italian astronomer Geminiano Montanari recorded observing variations in luminosity of the star Algol in 1667. Edmond Halley published the first measurements of the proper motion of a pair of nearby "fixed" stars, demonstrating that they had changed positions from the time of the ancient Greek astronomers Ptolemy and Hipparchus. The first direct measurement of the distance to a star (61 Cygni at 11.4 light-years) was made in 1838 by Friedrich Bessel using the parallax technique. Parallax measurements demonstrated the vast separation of the stars in the heavens.

William Herschel was the first astronomer to attempt to determine the distribution of stars in the sky. During the 1780s, he performed a series of gauges in 600 directions, and counted the stars observed along each line of sight. From this he deduced that the number of stars steadily increased toward one side of the sky, in the direction of the

Milky Way core. His son John Herschel repeated this study in the southern hemisphere and found a corresponding increase in the same direction. In addition to his other accomplishments, William Herschel is also noted for his discovery that some stars do not merely lie along the same line of sight, but are also physical companions that form binary star systems.

The science of stellar spectroscopy was pioneered by Joseph von Fraunhofer and Angelo Secchi. By comparing the spectra of stars such as Sirius to the Sun, they found differences in the strength and number of their absorption lines—the dark lines in a stellar spectra due to the absorption of specific frequencies by the atmosphere. In 1865 Secchi began classifying stars into spectral types. However, the modern version of the stellar classification scheme was developed by Annie J. Cannon during the 1900s.

Observation of double stars gained increasing importance during the nineteenth century. In 1834, Friedrich Bessel observed changes in the proper motion of the star Sirius, and inferred a hidden companion. Edward Pickering discovered the first spectroscopic binary in 1899 when he observed the periodic splitting of the spectral lines of the star Mizar in a 104 day period. Detailed observations of many binary star systems were collected by astronomers such as William Struve and S. W. Burnham, allowing the masses of stars to be determined from computation of the orbital elements. The first solution to the problem of deriving an orbit of binary stars from telescope observations was made by Felix Savary in 1827.

The twentieth century saw increasingly rapid advances in the scientific study of stars. The photograph became a valuable astronomical tool. Karl Schwarzschild discovered that the color of a star, and hence its temperature, could be determined by comparing the visual magnitude against the photographic magnitude. The development of the photoelectric photometer allowed very precise measurements of magnitude at multiple wavelength intervals. In 1921 Albert A. Michelson made the first measurements of a stellar diameter using an interferometer on the Hooker telescope.

Important conceptual work on the physical basis of stars occurred during the first decades of the twentieth century. In 1913, the Hertzsprung-Russell diagram was developed, propelling the astrophysical study of stars. Successful models were developed to explain the interiors of stars and stellar evolution. The spectra of stars were also successfully explained through advances in quantum physics. This allowed the chemical composition of the stellar atmosphere to be determined.

With the exception of supernovae, individual stars have primarily been observed in our Local Group of galaxies, and especially in the visible part of the Milky Way. But some stars have been observed in the M100 galaxy of the Virgo Cluster, about 100 million light years from the Earth. In the Local Supercluster it is possible to see star clusters, and current telescopes could in principle observe faint individual stars in the Local Cluster—the most distant stars resolved have up to hundred million light years away. However, outside the Local Supercluster of galaxies, few individual stars or clusters of stars have been observed. Exceptions include a faint image of a large star cluster containing hundreds of thousands of stars located one billion light years away—ten times the distance of the most distant star cluster previously observed; and an individual star at a distance of 9 billion light-years.

Multiverse

Multiverse is a hypothetical collection of potentially diverse observable universes, each of which would comprise everything that is experimentally accessible by a connected community of observers. The observable known universe, which is accessible to telescopes, is about 90 billion light-years across. However, this universe would constitute just a small or even infinitesimal subset of the multiverse. The multiverse idea has arisen in many versions, primarily in cosmology, quantum mechanics, and philosophy, and often asserts the actual physical existence of different potential configurations or histories of the known observable universe. The term multiverse was coined by American philosopher William James in 1895 to refer to the confusing moral meaning of natural phenomena and not to other possible universes.

Types of Multiverses

One useful way to classify multiverse models is by the degree to which the universes proposed by the model are connected—that is, by the degree to which they are part of a single system described by a well-defined physical and mathematical framework, generally with a common origin and possibly even interacting with one another.

On the fully disconnected end of this spectrum is the assertion that all possible worlds coexist with equal reality. This idea, known as modal realism, has been developed in philosophy, notably by the American David Kellogg Lewis in the 1970s and '80s. In physics and mathematics, meanwhile, it has been hypothesized (particularly in the 1990s by Swedish American physicist Max Tegmark and German computer scientist Jürgen Schmidhuber) that the known universe is equivalent to a mathematical formal system and that all such mathematical systems (or at least all of some class of such systems) are equally real. Similarly disconnected would be so-called parallel universes or other spiritual or religious planes of existence. Those other universes might be believed by some people to relate to the observable universe or even to interact with it, but exactly how these interactions would occur is not well defined.

Some what more-connected multiverses could arise from processes in quantum gravity, a hypothetical theory that would unite Einstein's theory of general relativity with quantum mechanics. General relativity describes how space-time evolves, but it does not answer the question of how a self-contained space-time could be created or destroyed.

One possible answer might be found in so-called cyclical cosmologies, where the "big crunch" of a collapsing universe would evolve into the big bang of a later expanding universe—presumably through quantum gravity or some other exotic process that is not described by well-understood physics. A related idea is that of the baby universe, in which a quantum gravitational process would create a new region of space-time that would bud off and potentially disconnect from its parent universe. This would lead to a "tree" of universes unlikely to interact after their formation. The process has been speculated to occur in the interiors of black holes.

The well-developed model of a multiverse of proliferating space-times is based on the idea of cosmological inflation. Inflation is a hypothetical process of the early universe in which space-time would have expanded exponentially at a much faster rate than at present. In most models, expansion would have been driven by energy present in the vacuum that would generate a repulsive force. This kind of exponential expansion creates a region of space-time far larger than the observable universe and leads to a highly connected (albeit fairly pedestrian) multiverse composed of regions very similar to the observable universe.

In an inflationary model, the transition to noninflation can occur at different times in different places. This leads to the fascinating phenomena that, in many versions of inflation, regions always exist where the transition to ordinary expansion has not yet taken place and where inflation is still occurring. This possibility leads to yet another picture in which inflation occurs eternally and generates an arbitrarily large or infinite number of post inflationary regions, one of which could be the observed universe.

Because the concept of inflation has both good theoretical justification and observational support and because the process of generating new universes through inflation is based on reasonably well-understood physics, this model of the multiverse has gained far more prominence than previous ideas. The inflationary multiverse is also fairly connected, in that all the universes would inhabit the same space-time and interactions between neighbouring universes might, in principle, produce observable effects. In the well-developed such model, the multiple universes are expanding bubbles in a common background; when bubbles collide, the resulting "bruise" could appear as a circular disturbance on the cosmic microwave background radiation.

Eternal inflation has an interesting synergy with string theory. In string theory a compact space of six or more dimensions is associated with each point in traditional four-dimensional space-time. (Thus, the universe actually has 10 or more dimensions.) The attributes of these compact spaces can determine physical conditions throughout the rest of the universe. The compact space's configuration also determines a vacuum energy, which if positive will generally drive eternal inflation. Moreover, transitions between configurations can occur, leading to multiple universes with different physical constants and fields. This sort of multiverse has been dubbed "eternal inflation/string landscape." Because physical constants would take many values in many actual universes, this framework has been proposed as a so-called anthropic explanation for why certain physical constants—most notably the cosmological constant—take a value compatible with a universe suited for life but very different from a "natural" value. In the eternal inflation/string landscape model, much of the multiverse would be uninhabited, and life necessarily would be found only in a region with compatible physical constants.

A rather different but important type of multiverse potentially arises directly from quantum mechanics. According to quantum mechanics, a system can be described by states corresponding to different outcomes of a physical measurement of the system by a macroscopic apparatus. Before a system has been measured, it will be in a superposition of such states (mathematically, a sum of state vectors weighted by complex coefficients). Such a system does not admit a definite prediction for a measurement outcome but allows only probabilities for outcomes. An open question is how these probabilities should be regarded. In particular, in an isolated system the evolution of the state vector is deterministic, whereas during a measurement it appears to jump indeterministically to the state corresponding to the measured value. American physicist Hugh Everett proposed that this is an artifact of treating the measuring apparatus classically. He proposed instead that the apparatus (and any observers) interacting with a quantum system simply "enlarge" the superposition to include a superposition between macroscopic states in which different experimental outcomes occur. Thus, all possible outcomes occur but in different worlds. Often termed the "many worlds" interpretation of quantum mechanics, this sort of multiverse is very connected in that it is mathematically unified. However, calculations of the decoherence between universes indicates that it is essentially impossible for macroscopic objects in separate universes to interact.

Finally, it should be noted that different types of multiverses may combine or coexist. For example, an Everett-type multiverse could include different worlds, each of which is a different eternally inflating multiverse, or eternal inflation could, in principle, spawn multiple cyclic multiverses or result from the creation of a baby universe.

Critiques of Multiverse Theories

Multiverse theories have been widely criticized as speculation or philosophy rather than science, which is indeed a legitimate concern regarding the more-disconnected types of multiverses. Less legitimate are criticisms of multiverse models that appear as a side effect of well-motivated and tested theories or that admit some level of empirical verification.

Some of the fiercest debate surrounds the legitimacy and utility of anthropic arguments such as that which explains the small observed value of the cosmological constant. On the one hand, such reasoning is arguably no different from explaining the highly unusual ambient temperature and local density experienced on Earth relative to most locations in the universe because only a certain range of temperatures and densities can support life. On the other hand, it is very difficult to define what life exactly is, or what it requires, in a multiverse context. It is also very difficult to compute or compare probabilities for a given sort of life to arise in different sorts of universes. There is also a danger of accepting anthropic explanations for things that could be better and more elegantly explained using other reasoning.

It is also debatable whether multiverses are in accord with Ockham's razor, which holds that the simplest explanation of an entity or phenomenon is the one that should be preferred. Critics contend that other universes are almost perfect examples of entities "multiplied beyond necessity." Multiverse defenders, on the other hand, contend that a description of the observable universe as one part of a multiverse can be much simpler than a description of the observable universe as the only universe at all.

Friedmann Universe

Friedmann Universe model was developed in 1922 by the Russian meteorologist and mathematician Aleksandr Friedmann. He believed that Albert Einstein's general theory of relativity required a theory of the universe in motion, as opposed to the static universe that scientists until then had proposed. He hypothesized a big bang followed by expansion, then contraction and an eventual big crunch. This model supposes a closed universe, but he also proposed similar solutions involving an open universe (which expands infinitely) or a flat universe (in which expansion continues infinitely but gradually approaches a rate of zero).

2

Cosmology: A Comprehensive Study

The branch of astronomy that deals with the study of origin and evolution of universe is termed as cosmology. Dualistic cosmology involves diverse views from different traditional and scriptural religions. The topics elaborated in this chapter will help in gaining a better perspective of cosmology.

Cosmology is the study of the nature of the universe as a whole entity. Cosmologists are interested in the formation, evolution and future of the universe and its constituents.

Most objects we can see with telescopes are large or exist at extreme distances (e.g. planets, stars, galaxies, clusters of galaxies and even superclusters). The majority view of cosmologists is that all these objects were formed after an initial, extremely hot and dense formation event, about 14 Gigayears ago, that has created (and continues to create) the space we see around us. This event is called the Big Bang.

Whilst the hot Big Bang model does seem to explain much of what we observe around us, there are still many fundamental questions that exist. What is the majority of the matter in the universe made of? How common are planets around stars? What causes some galaxies to be elliptical, spiral or irregular in shape? What is the geometry of the universe? What is the mysterious dark energy? Is there a cosmological constant? Is it a variable? Do other universes exist?

As well as the properties of the largest objects (e.g. galaxies and large scale structure), cosmology is becoming increasingly concerned with the properties of the smallest objects.

To help determine what happened at the beginning of the universe, cosmologists need the help of particle physicists. The Big Bang model describes a very hot and dense beginning to the universe in which many interesting particle physics phenomena occur. These phenomena have influenced the type of universe we live in.

In the earliest stages the universe was tremendously hot and matter could not exist. The universe was radiation dominated. As the universe expanded and cooled elementary particles could be created, which later formed the lightest elements such as hydrogen, helium and lithium. Heavier elements had to wait for stars to form so that they could be made via nucleosynthesis in the high temperature, pressure and density centres of massive stars.

The Standard Model of particle physics is a mathematical description of the 12 fundamental particles (6 leptons and 6 quarks) and 3 forces (electromagnetic, weak and strong). It is thought that at ~10^{-11} seconds after the Big Bang all 4 (current epoch) forces (the three mentioned above plus gravity) became separate forces. However at around ~10^{-43} seconds after the Big Bang (the Planck Time) all 4 forces were unified into one single force. The process of the forces separating from each other is called spontaneous symmetry breaking.

The first cosmologists were Babylonians and Egyptians who observed the sky and who could predict the apparent motions of the Sun, moon, brightest stars and the planets.

In the 4th century BC, Greek philosophers deduced that the stars were fixed on a celestial sphere which rotated about the spherical Earth. The planets, Sun and the Moon moved in a fluid substance called ether between the Earth and the stars.

Aristotle's spherical cosmology.

In the 2nd century AD Ptolemy based his work on the belief that all motion was circular. To account for the motion of some of the planets, which seem to loop back upon themselves, Ptolemy introduced epicycles so that the planets moved in circles upon circles.

New observations drive advances in theory, and new theories can spur new observations. However many centuries passed until a significant new development in cosmology occured.

In the 16th century Nicholas Copernicus proposed a heliocentric system in which the Earth rotated on its axis, and along with the other planets, orbited the Sun. But the observational evidence of the time favoured the epicycle-based Ptolemaic system. The Copernican system was promoted by some but it was the discovery of the aberration of starlight in 1728 that proved without doubt that the Earth orbits the Sun.

In the early 17th century Galileo Galilei discovered moons orbiting the planet Jupiter. It clearly showed that Earth was not special and made many believe in the Copernican heliocentric model of planets orbiting the Sun. Isaac Newton then discovered the inverse-square law for the gravitational force which could explain the elliptical orbits of planets and comets in the Solar System. A physical framework for celestial motions had been found.

If the Earth orbited the Sun then positions of nearby stars, compared to the background, should change. However, initial observations did not detect any such motion. The absence of any observable shift or parallax in the positions of the stars as the Earth orbited the Sun implied that stars must be at great distances from the Sun. Newton concluded that the universe must be an infinite and eternal sea of stars, each much like our own Sun.

In the 18th century two notable philosophers emerged with similar ideas. In 1750 Thomas Wright suggested that the Milky Way, the Galaxy, was a vast spinning disk consisting of stars and planets. Immanuel Kant wrote the "General Natural History and Theory of the Heavens" in 1755 in which he suggested that the spiral nebulae, faint nebulous objects observed across the sky, were external galaxies or island universes independent of the Milky Way.

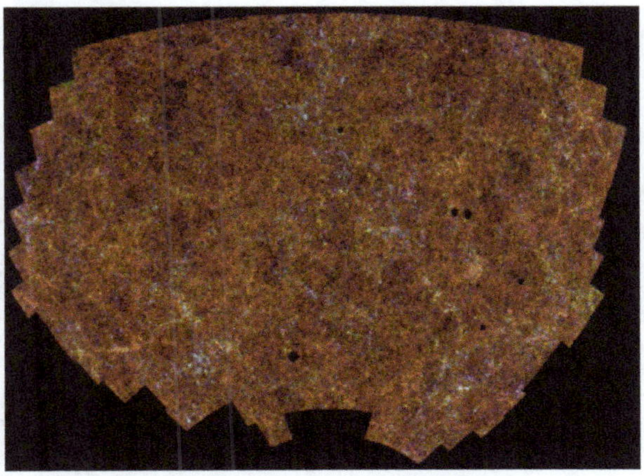

3 million galaxies across 7000 square degrees of sky.

Physical cosmology, the quantitative version of cosmology, began with Albert Einstein in 1915 when he developed the first substantial models of the universe via the solutions to his General Theory of Relativity. These solutions were added to and improved upon by Alexander Freidmann, Willem de Sitter, Georges Lemaitre, H. P. Robertson and

Arthur Geoffrey Walker. At that stage astronomers were not aware of the expansion of the universe, and Einstein had introduced a mathematical term, a cosmological constant, to ensure his universe was static.

In 1912 Henrietta Leavitt discovered Cepheid variable stars in the Magellanic Clouds and confirmed that the variables with longer periods had larger luminosities. From 1912 onwards Vesto Slipher at Lowell Observatory began to accumulate velocities of spiral nebulae. By the mid-1920s the vast majority of these nebulae had velocities of recession, some in the thousands of km/s. It was difficult not to assign such extreme velocities to external galaxies.

By the mid 1920s J. C. Duncan and Edwin Hubble had also detected Cepheid variable stars in Messier 33, Messier 31 and NGC 6822. These stars obeyed a Period-Luminosity relationship in which the period of their variability was related to their intrinsic brightness. Once a calibration of nearby Cepheids with known distances had been established a distance to each variable could be assigned based solely on it's period of variability. Based on these variable stars the inferred distances were far too great to allow these spiral nebulae to be part of our Galaxy. They were galaxies in their own right. Wright and Kant were correct.

In 1929 Hubble published a landmark paper that described a relation between the distance to a galaxy and its observed radial velocity. More distant galaxies have larger recessional velocities (as had been seen with a smaller sample by Vesto Slipher). A case can be made that Lundmark and Lemaitre had beaten Hubble to this discovery. As well, H.P. Robertson was the first to describe the results as cosmic expansion. Einstein then threw away his cosmological constant. The universe was not static, but expanding.

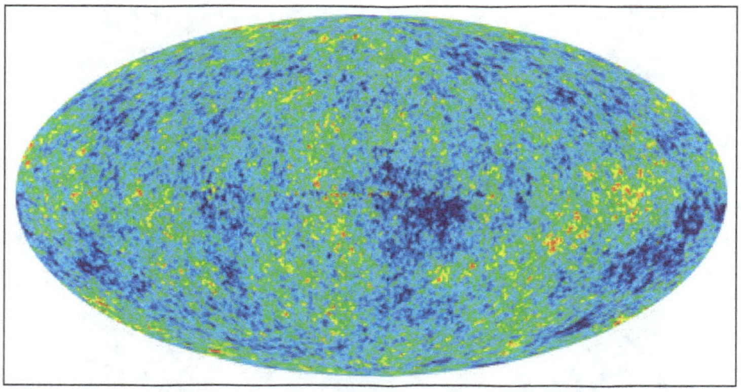

WMAP views the cosmic microwave background.

In the 1950s Hermann Bondi, Thomas Gold and Fred Hoyle put forward the Steady State model of the universe. In this model, matter was made from the vacuum of space and the universe looked the same in any direction and at any time. Philosophically elegant, it avoided a start and finish to the universe. It was, however, testable. The observed evolution of radio sources, and later the observed peak in number density of quasars, several billion light years distant, called the Steady State model into question.

In the early 1960s Arno Penzias and Robert Wilson detected a very faint signal in the microwave region corresponding to all-sky radiation at ~3 degrees K that peaks in intensity at a wavelength of 2 mm. Robert Dicke and collaborators had predicted such a signal, the Cosmic Background, if the universe had begun in a hot, dense state.

The 3 degrees K radiation that we now detect was emitted about 300,000 years after the Big Bang when the universe had cooled enough for plasma to form a gas of neutral atoms. After that the photons of the cosmic background radiation have traveled on straight lines (to us) without interacting with matter ever since. Further observations showed that the radiation was of the expected blackbody form for a very hot origin and the hot Big Bang model was re-kindled as the preferred cosmological model. Along with the Cosmic Background, two other observations underpin the Big Bang model. One is that nucleosynthesis in an early hot universe correctly accounts for the cosmic abundance of the light nuclear isotopes such as hydrogen, deuterium, helium-3, helium-4, and lithium-7. The second was the observed expansion of the universe as implied by the motions of distant galaxies.

However, the Big Bang model could not explain certain observations of the universe. If we divide space into cubes of several hundred million light-years, each such cube would look similar in terms of mass density, galaxy density and the amount of coherent structure. This large-scale uniformity is observed in distant galaxy surveys. However, in the standard Big Bang model the universe evolves so quickly that there is no time for such similarity to be established. This problem is known as the horizon problem where horizon is used to indicate the largest distance that information or energy could have traversed since the instant of the Big Bang, given the restriction of the constancy of the speed of light. Simply put, the universe is nearly homogenous and isotropic on very large scales.

The second problem is the flatness problem. The mass density of the universe governs its evolution and fate. If the mass density exceeds a critical density, then gravity will be strong enough to reverse the current expansion and the universe has a geometry which is called closed. If the mass density is less than the critical value, the universe will go on expanding forever, and the universe has an open geometry. The ratio of the actual mass density to the critical value is known as Omega. The General Theory of Relativity implies that the geometry of the universe is Euclidean only if Omega is exactly 1.0, so an Omega = 1 universe is called flat. At present time we believe the value of Omega to be well within a factor of 10 of 1.0. For the value of Omega to be this near to 1.0 at this epoch is remarkable – if it was initially only a very, small amount away from 1.0, some 14 billion years of evolution would have driven it well away from this value at this current epoch. To all purposes the universe appears flat.

Three Types of Geometry of the Universe

To explain these observed 'problems', in 1980 Alan Guth determined that a period of extremely rapid exponential expansion, 'inflation', occurred around 10-34 seconds after the Big Bang. Immediately after this inflationary period, the universe currently visible to us had a radius of ~1 m. The universe then returned to its normal (linear) mode of expansion. Subsequent fine tuning of inflation has been done by Andrei Linde, Andreas Albrecht and Paul Steinhardt, and current versions include multi-universes all undergoing some type of inflation (e.g. chaotic, eternal). Inflation may be linked with string theory and brane cosmology in which our 4 dimensional (3 spatial, one time dimension) brane can be considered as a subset of a much higher dimension bulk.

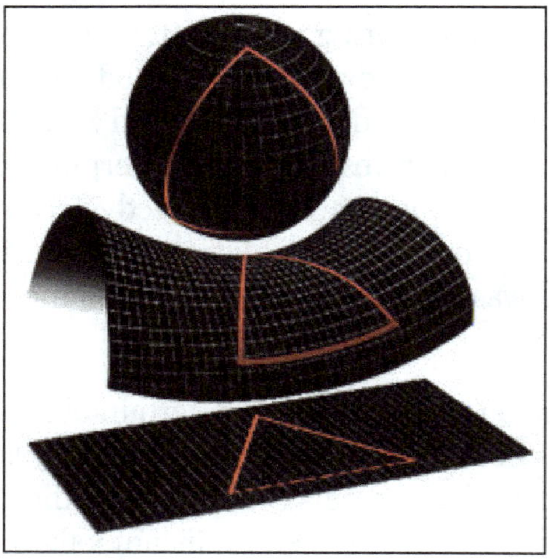

Three basic geometrical forms of the universe. From top: closed, open and flat.

The very early universe inflation period, and its physical causes, may be related to a recent observation of the universe. Distant high redshift supernovae appear to be slightly less luminous than expected, which can be interpreted as a universe that is undergoing a phase of accelerated expansion. Dark energy has been postulated as a sort of anti-gravity which drives this acceleration.

Current topical areas in cosmological research include the nature of dark matter and dark energy, the search for the first stars and galaxies, properties of the cosmic background, and numerical simulations of large scale structure.

Relativistic Cosmologies

Einstein's Model

To derive his 1917 cosmological model, Einstein made three assumptions that lay outside the scope of his equations. The first was to suppose that the universe is homogeneous and isotropic in the large (i.e., the same everywhere on average at any instant in time), an assumption that the English astrophysicist Edward A. Milne later elevated to an entire philosophical outlook by naming it the cosmological principle. Given the success of the Copernican revolution, this outlook is a natural one. Newton himself had it implicitly in mind when he took the initial state of the universe to be everywhere the same before it developed "ye Sun and Fixt stars."

The second assumption was to suppose that this homogeneous and isotropic universe had a closed spatial geometry. The total volume of a three-dimensional space with uniform positive curvature would be finite but possess no edges or boundaries (to be consistent with the first assumption).

The third assumption made by Einstein was that the universe as a whole is static—i.e., its large-scale properties do not vary with time. This assumption, made before Hubble's observational discovery of the expansion of the universe, was also natural; it was the simplest approach, as Aristotle had discovered, if one wishes to avoid a discussion of a creation event. Indeed, the philosophical attraction of the notion that the universe on average is not only homogeneous and isotropic in space but also constant in time was so appealing that a school of English cosmologists—Hermann Bondi, Fred Hoyle, and Thomas Gold—would call it the perfect cosmological principle and carry its implications in the 1950s to the ultimate refinement in the so-called steady-state theory.

To his great chagrin Einstein found in 1917 that with his three adopted assumptions, his equations of general relativity—as originally written down—had no meaningful solutions. To obtain a solution, Einstein realized that he had to add to his equations an extra term, which came to be called the cosmological constant. If one speaks in Newtonian terms, the cosmological constant could be interpreted as a repulsive force of unknown origin that could exactly balance the attraction of gravitation of all the matter in Einstein's closed universe and keep it from moving. The inclusion of such a term in a more general context, however, meant that the universe in the absence of any mass-energy (i.e., consisting of a vacuum) would not have a space-time structure that was flat (i.e., would not have satisfied the dictates of special relativity exactly). Einstein was prepared to make such a sacrifice only very reluctantly, and, when he later learned of Hubble's discovery of the expansion of the universe and realized that he could have predicted it had he only had more faith in the original form of his equations, he regretted the introduction of the cosmological constant as the "biggest blunder" of his life. Ironically, observations of distant supernovas have shown the existence of dark energy, a repulsive force that is the dominant component of the universe.

De Sitter's Model

It was also in 1917 that the Dutch astronomer Willem de Sitter recognized that he could obtain a static cosmological model differing from Einstein's simply by removing all matter. The solution remains stationary essentially because there is no matter to move about. If some test particles are reintroduced into the model, the cosmological term would propel them away from each other. Astronomers now began to wonder if this effect might not underlie the recession of the spiral galaxies.

Friedmann-Lemaître Models

In 1922 Aleksandr A. Friedmann, a Russian meteorologist and mathematician, and in

1927 Georges Lemaître, a Belgian cleric, independently discovered solutions to Einstein's equations that contained realistic amounts of matter. These evolutionary models correspond to big bang cosmologies. Friedmann and Lemaître adopted Einstein's assumption of spatial homogeneity and isotropy (the cosmological principle). They rejected, however, his assumption of time independence and considered both positively curved spaces ("closed" universes) as well as negatively curved spaces ("open" universes). The difference between the approaches of Friedmann and Lemaître is that the former set the cosmological constant equal to zero, whereas the latter retained the possibility that it might have a nonzero value.

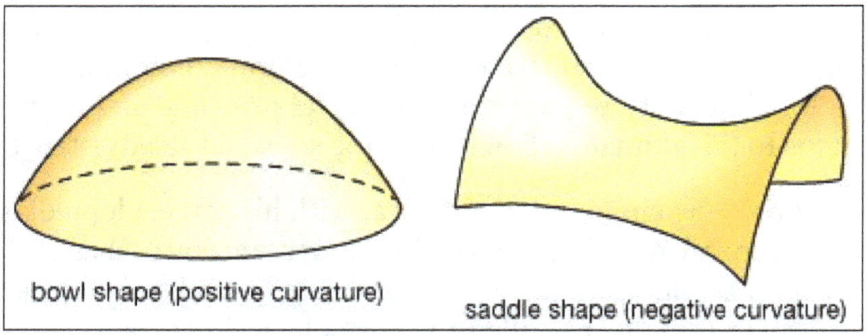

Intrinsic curvature of a surface.

The decision to abandon a static model meant that the Friedmann models evolve with time. As such, neighbouring pieces of matter have recessional (or contractional) phases when they separate from (or approach) one another with an apparent velocity that increases linearly with increasing distance. Friedmann's models thus anticipated Hubble's law before it had been formulated on an observational basis. It was Lemaître, however, who had the good fortune of deriving the results at the time when the recession of the galaxies was being recognized as a fundamental cosmological observation, and it was he who clarified the theoretical basis for the phenomenon.

The geometry of space in Friedmann's closed models is similar to that of Einstein's original model; however, there is a curvature to time as well as one to space. Unlike Einstein's model, where time runs eternally at each spatial point on an uninterrupted horizontal line that extends infinitely into the past and future, there is a beginning and end to time in Friedmann's version of a closed universe when material expands from or is recompressed to infinite densities. These instants are called the instants of the "big bang" and the "big squeeze," respectively. The global space-time diagram for the middle half of the expansion-compression phases can be depicted as a barrel lying on its side. The space axis corresponds again to any one direction in the universe, and it wraps around the barrel. Through each spatial point runs a time axis that extends along the length of the barrel on its (space-time) surface. Because the barrel is curved in both space and time, the little squares in the grid of the curved sheet of graph paper marking the space-time surface are of nonuniform size, stretching to become bigger when the barrel broadens (universe expands) and shrinking to become smaller when the barrel narrows (universe contracts).

It should be remembered that only the surface of the barrel has physical significance; the dimension off the surface toward the axle of the barrel represents the fourth spatial dimension, which is not part of the real three-dimensional world. The space axis circles the barrel and closes upon itself after traversing a circumference equal to $2\pi R$, where R, the radius of the universe (in the fourth dimension), is now a function of the time t. In a closed Friedmann model, R starts equal to zero at time t = 0 (not shown in barrel diagram), expands to a maximum value at time $t = t_m$ (the middle of the barrel), and recontracts to zero at time $t = 2t_m$, with the value of tm dependent on the total amount of mass that exists in the universe.

Imagine now that galaxies reside on equally spaced tick marks along the space axis. Each galaxy on average does not move spatially with respect to its tick mark in the spatial (ringed) direction but is carried forward horizontally by the march of time. The total number of galaxies on the spatial ring is conserved as time changes, and therefore their average spacing increases or decreases as the total circumference $2\pi R$ on the ring increases or decreases (during the expansion or contraction phases). Thus, without in a sense actually moving in the spatial direction, galaxies can be carried apart by the expansion of space itself. From this point of view, the recession of galaxies is not a "velocity" in the usual sense of the word. For example, in a closed Friedmann model, there could be galaxies that started, when R was small, very close to the Milky Way system on the opposite side of the universe. Now, 10^{10} years later, they are still on the opposite side of the universe but at a distance much greater than 10^{10} light-years away. They reached those distances without ever having had to move (relative to any local observer) at speeds faster than light—indeed, in a sense without having had to move at all. The separation rate of nearby galaxies can be thought of as a velocity without confusion in the sense of Hubble's law, if one wants, but only if the inferred velocity is much less than the speed of light.

On the other hand, if the recession of the galaxies is not viewed in terms of a velocity, then the cosmological redshift cannot be viewed as a Doppler shift. How, then, does it arise? The answer is contained in the barrel diagram when one notices that, as the universe expands, each small cell in the space-time grid also expands. Consider the propagation of electromagnetic radiation whose wavelength initially spans exactly one cell length, so that its head lies at a vertex and its tail at one vertex back. Suppose an elliptical galaxy emits such a wave at some time t_1. The head of the wave propagates from corner to corner on the little square grids that look locally flat, and the tail propagates from corner to corner one vertex back. At a later time t_2, a spiral galaxy begins to intercept the head of the wave. At time t_2, the tail is still one vertex back, and therefore the wave train, still containing one wavelength, now spans one current spatial grid spacing. In other words, the wavelength has grown in direct proportion to the linear expansion factor of the universe. Since the same conclusion would have held if n wavelengths had been involved instead of one, all electromagnetic radiation from a given object will show the same cosmological redshift if the universe (or, equivalently, the average spacing between galaxies) was smaller at the epoch of transmission than at the epoch of reception. Each wavelength will have been stretched in direct proportion to the expansion of the universe in between.

A nonzero peculiar velocity for an emitting galaxy with respect to its local cosmological frame can be taken into account by Doppler-shifting the emitted photons before applying the cosmological redshift factor; i.e., the observed redshift would be a product of two factors. When the observed redshift is large, one usually assumes that the dominant contribution is of cosmological origin. When this assumption is valid, the redshift is a monotonic function of both distance and time during the expansional phase of any cosmological model. Thus, astronomers often use the redshift z as a shorthand indicator of both distance and elapsed time. Following from this, the statement "object X lies at z = a" means that "object X lies at a distance associated with redshift a"; the statement "event Y occurred at redshift z = b" means that "event Y occurred a time ago associated with redshift b."

The open Friedmann models differ from the closed models in both spatial and temporal behaviour. In an open universe the total volume of space and the number of galaxies contained in it are infinite. The three-dimensional spatial geometry is one of uniform negative curvature in the sense that, if circles are drawn with very large lengths of string, the ratio of circumferences to lengths of string are greater than 2π. The temporal history begins again with expansion from a big bang of infinite density, but now the expansion continues indefinitely, and the average density of matter and radiation in the universe would eventually become vanishingly small. Time in such a model has a beginning but no end.

Einstein–De Sitter Universe

In 1932 Einstein and de Sitter proposed that the cosmological constant should be set equal to zero, and they derived a homogeneous and isotropic model that provides the separating case between the closed and open Friedmann models; i.e., Einstein and de Sitter assumed that the spatial curvature of the universe is neither positive nor negative but rather zero. The spatial geometry of the Einstein–de Sitter universe is Euclidean (infinite total volume), but space-time is not globally flat (i.e., not exactly the space-time of special relativity). Time again commences with a big bang and the galaxies recede forever, but the recession rate (Hubble's "constant") asymptotically coasts to zero as time advances to infinity. Because the geometry of space and the gross evolutionary properties are uniquely defined in the Einstein–de Sitter model, many people with a philosophical bent long considered it the most fitting candidate to describe the actual universe.

Bound and Unbound Universes and the Closure Density

The different separation behaviours of galaxies at large timescales in the Friedmann closed and open models and the Einstein–de Sitter model allow a different classification scheme than one based on the global structure of space-time. The alternative way of looking at things is in terms of gravitationally bound and unbound systems: closed models where galaxies initially separate but later come back together again represent bound universes; open models where galaxies continue to separate forever represent unbound universes; the Einstein–de Sitter model where galaxies separate forever but slow to a halt at infinite time represents the critical case.

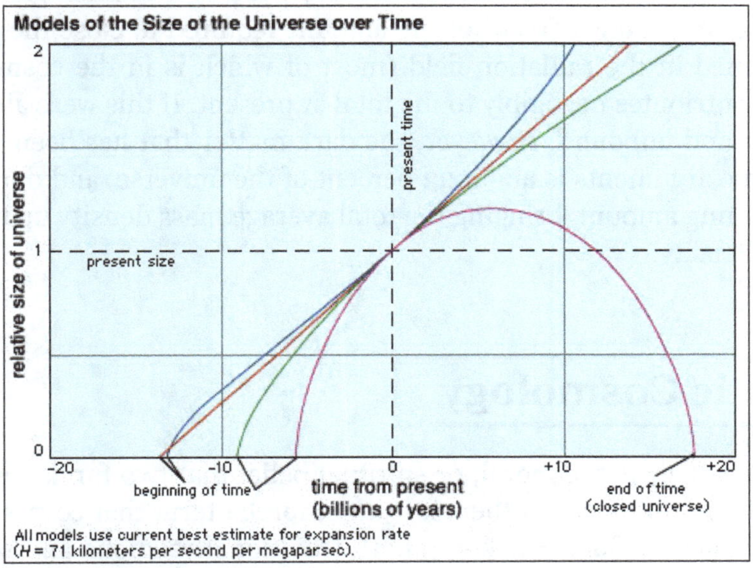

How the relative size of the universe changes with time in four different models. The red line shows a universe devoid of matter, with constant expansion. Pink shows a collapsing universe, with six times the critical density of matter. Green shows a model favoured until 1998, with exactly the critical density and a universe 100 percent matter. Blue shows the currently favoured scenario, with exactly the critical density, of which 27 percent is visible and dark matter and 73 percent is dark energy.

The advantage of this alternative view is that it focuses attention on local quantities where it is possible to think in the simpler terms of Newtonian physics—attractive forces, for example. It is intuitively clear that the feature that should distinguish whether or not gravity is capable of bringing a given expansion rate to a halt depends on the amount of mass (per unit volume) present. This is indeed the case; the Newtonian and relativistic formalisms give the same criterion for the critical, or closure, density (in mass equivalent of matter and radiation) that separates closed or bound universes from open or unbound ones. If Hubble's constant at the present epoch is denoted as H_0, then the closure density (corresponding to an Einstein–de Sitter model) equals $3H_0^2/8\pi G$, where G is the universal gravitational constant in both Newton's and Einstein's theories of gravity. The numerical value of Hubble's constant H_0 is 22 kilometres per second per million light-years; the closure density then equals 10^{-29} gram per cubic centimetre, the equivalent of about six hydrogen atoms on average per cubic metre of cosmic space. If the actual cosmic average is greater than this value, the universe is bound (closed) and, though currently expanding, will end in a crush of unimaginable proportion. If it is less, the universe is unbound (open) and will expand forever. The result is intuitively plausible since the smaller the mass density, the smaller the role for gravitation, so the more the universe will approach free expansion (assuming that the cosmological constant is zero).

The mass in galaxies observed directly, when averaged over cosmological distances, is estimated to be only a few percent of the amount required to close the universe. The amount contained in the radiation field (most of which is in the cosmic microwave background) contributes negligibly to the total at present. If this were all, the universe would be open and unbound. However, the dark matter that has been deduced from various dynamic arguments is about 23 percent of the universe, and dark energy supplies the remaining amount, bringing the total average mass density up to 100 percent of the closure density.

Dualistic Cosmology

Dualism in cosmology is the moral, or spiritual belief that two fundamental concepts exist, which often oppose each other. It is an umbrella term that covers a diversity of views from various religions, including both traditional religions and scriptural religions.

Moral dualism is the belief of the great complement of, or conflict between, the benevolent and the malevolent. It simply implies that there are two moral opposites at work, independent of any interpretation of what might be "moral" and independent of how these may be represented. Moral opposites might, for example, exist in a worldview which has one god, more than one god, or none. By contrast, duotheism, bitheism or ditheism implies (at least) two gods. While bitheism implies harmony, ditheism implies rivalry and opposition, such as between good and evil, or light and dark, or summer and winter. For example, a ditheistic system could be one in which one god is a creator, and the other a destroyer. In theology, dualism can also refer to the relationship between the deity and creation or the deity and the universe (see theistic dualism). This form of dualism is a belief shared in certain traditions of Christianity and Hinduism. Alternatively, in ontological dualism, the world is divided into two overarching categories. The opposition and combination of the universe's two basic principles of yin and yang is a large part of Chinese philosophy, and is an important feature of Taoism. It is also discussed in Confucianism.

Many myths and creation motifs with dualistic cosmologies have been described in ethnographic and anthropological literature. These motifs conceive the world as being created, organized, or influenced by two demiurges, culture heroes, or other mythological beings, who either compete with each other or have a complementary function in creating, arranging or influencing the world. There is a huge diversity of such cosmologies. In some cases, such as among the Chukchi, the beings collaborate rather than competing, and contribute to the creation in a coequal way. In many other instances the two beings are not of the same importance or power (sometimes, one of them is even characterized as gullible). Sometimes they can be contrasted as good versus evil. They may be often believed to be twins or at least brothers. Dualistic motifs in mythologies can be observed in all inhabited continents. Zolotaryov concludes that they cannot be explained by diffusion or borrowing, but are rather of convergent origin: they are related to a dualistic organization of society (moieties); in some cultures, this social organization may have ceased to exist, but mythology preserves the memory in more and more disguised ways.

Moral Dualism

Moral dualism is the belief of the great complement or conflict between the benevolent and the malevolent. Like ditheism/bitheism (see below), moral dualism does not imply the absence of monist or monotheistic principles. Moral dualism simply implies that there are two moral opposites at work, independent of any interpretation of what might be "moral" and—unlike ditheism/bitheism—independent of how these may be represented.

For example, Mazdaism (Mazdean Zoroastrianism) is both dualistic and monotheistic (but not monist by definition) since in that philosophy God—the Creator—is purely good, and the antithesis—which is also uncreated–is an absolute one. Zurvanism (Zurvanite Zoroastrianism), Manichaeism, and Mandaeism are representative of dualistic *and* monist philosophies since each has a supreme and transcendental First Principle from which the two equal-but-opposite entities then emanate. This is also true for the lesser-known Christian gnostic religions, such as Bogomils, Catharism, and so on. More complex forms of monist dualism also exist, for instance in Hermeticism, where *Nous* "thought"—that is described to have created man—brings forth both good and evil, dependent on interpretation, whether it receives prompting from the God or from the Demon. Duality with pluralism is considered a logical fallacy.

Duotheism, Bitheism and Ditheism

When used with regards to multiple gods, *dualism* may refer to duotheism, bitheism, or ditheism. Although ditheism/bitheism imply moral dualism, they are not equivalent: ditheism/bitheism implies (at least) two gods, while moral dualism does not necessarily imply *theism* (theos = god) at all.

Both bitheism and ditheism imply a belief in two equally powerful gods with complementary or antonymous properties; however, while bitheism implies harmony, ditheism implies rivalry and opposition, such as between good and evil, bright and dark, or summer and winter. For example, a ditheistic system would be one in which one god is creative, the other is destructive (cf. theodicy). In the original conception of Zoroastrianism, for example, Ahura Mazda was the spirit of ultimate good, while Ahriman (Angra Mainyu) was the spirit of ultimate evil.

In a bitheistic system, by contrast, where the two deities are not in conflict or opposition, one could be male and the other female (cf. duotheism). One well-known example of a bitheistic or duotheistic theology based on gender polarity is found in the neopagan religion of Wicca. In Wicca, dualism is represented in the belief of a god and a goddess as a dual partnership in ruling the universe. This is centered on the worship of a divine couple, the Moon Goddess and the Horned God, who are regarded as lovers. However, there is also a ditheistic theme within traditional Wicca, as the Horned God has dual aspects of bright and dark - relating to day/night, summer/winter - expressed as the Oak King and the Holly King, who in Wiccan myth and ritual are said to engage in battle twice a year for the hand of the Goddess, resulting in the changing seasons. (Within Wicca, bright and dark do not correspond to notions of "good" and "evil" but are aspects of the natural world, much like yin and yang in Taoism.)

Radical and Mitigated Dualism

- Radical Dualism – or absolute Dualism which posits two co-equal divine forces. Manichaeism conceives of two previously coexistent realms of light and darkness which become embroiled in conflict, owing to the chaotic actions of the latter. Subsequently, certain elements of the light became entrapped within darkness; the purpose of material creation is to enact the slow process of extraction of these individual elements, at the end of which the kingdom of light will prevail over darkness. Manicheanism likely inherits this dualistic mythology from Zoroastrianism, in which the eternal spirit Ahura Mazda is opposed by his antithesis, Angra Mainyu; the two are engaged in a cosmic struggle, the conclusion of which will likewise see Ahura Mazda triumphant. 'The Hymn of the Pearl' included the belief that the material world corresponds to some sort of malevolent intoxication brought about by the powers of darkness to keep elements of the light trapped inside it in a state of drunken distraction.

- Mitigated Dualism – is where one of the two principles is in some way inferior to the other. Such classical Gnostic movements as the Sethians conceived of the material world as being created by a lesser divinity than the true God that was the object of their devotion. The spiritual world is conceived of as being radically different from the material world, co-extensive with the true God, and the true home of certain enlightened members of humanity; thus, these systems were expressive of a feeling of acute alienation within the world, and their resultant aim was to allow the soul to escape the constraints presented by the physical realm.

However, bitheistic and ditheistic principles are not always so easily contrastable, for instance in a system where one god is the representative of summer and drought and the other of winter and rain/fertility (cf. the mythology of Persephone). Marcionism, an early Christian sect, held that the Old and New Testaments were the work of two opposing gods: both were First Principles, but of different religions.

Theistic Dualism

In theology, dualism can refer to the relationship between God and creation or God and the universe. This form of dualism is a belief shared in certain traditions of Christianity and Hinduism.

The dualism between God and Creation has existed as a central belief in multiple historical sects and traditions of Christianity, including Marcionism, Catharism, Paulicianism, and Gnostic Christianity. Christian dualism refers to the belief that God and creation are distinct, but interrelated through an indivisible bond. In sects like the Cathars and the Paulicians, this is a dualism between the material world, created by an evil god, and a moral god. Historians divide Christian dualism into absolute dualism, which held that the good and evil gods were equally powerful, and mitigated dualism, which held that material evil was subordinate to the spiritual good. The belief, by Christian theologians who adhere to a libertarian or compatibilist view of free will, that free will separates humankind from God has also been characterized as a form of dualism. The theologian Leroy Stephens Rouner compares the dualism of Christianity with the dualism that exists in Zoroastrianism and the Samkhya tradition of Hinduism. The theological use of the word dualism dates back to 1700, in a book that describes the dualism between good and evil.

The tolerance of dualism ranges widely among the different Christian traditions. As a monotheistic religion, the conflict between dualism and monism has existed in Christianity since its inception. In the Catholic Church, "the dualistic hypothesis of an eternal world existing side by side with God was of course rejected" by the thirteenth century, but mind–body dualism was not. The problem of evil is difficult to reconcile with absolute monism, and has prompted some Christian sects to veer towards dualism. Gnostic forms of Christianity were more dualistic, and some Gnostic traditions posited that the Devil was separate from God as an independent deity. The Christian dualists of the Byzantine Empire, the Paulicians, were seen as Manichean heretics by Byzantine theologians. This tradition of Christian dualism, founded by Constantine-Silvanus, argued that the universe was created through evil and separate from a moral God.

The Cathars, a Christian sect in southern France, believed that there was a dualism between two gods, one representing good and the other representing evil. The Roman Catholic Church denounced the Cathars as heretics, and sought to crush the movement in the 13th century. The Albigensian Crusade was initiated by Pope Innocent III in 1208 to remove the Cathars from Languedoc in France, where they were known as Albigesians. The Inquisition, which began in 1233 under Pope Gregory IX, also targeted the Cathars.

Gnosticism

Gnosticism is a diverse, syncretistic religious movement consisting of various belief systems generally united in the teaching that humans are divine souls trapped in a material world created by an imperfect god, the demiurge. The demiurge may be depicted as an embodiment of evil, or in other instances as merely imperfect and as benevolent as its inadequacy permits. This demiurge exists alongside another remote and unknowable supreme being that embodies good.

Bogomils, Paulicians and Cathars are typically seen as being imitative of Gnosticism. Whether or not the Cathari possessed direct historical influence from ancient Gnosticism is a matter of dispute. The basic conceptions of Gnostic cosmology are, however, to be found in Cathar beliefs (most distinctly in their notion of a lesser creator god). Unlike the second century Gnostics, they did not apparently place any special relevance upon knowledge (*gnosis*) as an effective salvific force.

Ontological Dualism

Alternatively, *dualism* can mean the tendency of humans to perceive and understand the world as being divided into two overarching categories. In this sense, it is dualistic when one perceives a tree as a thing separate from everything surrounding it. This form of ontological dualism exists in Taoism and Confucianism, beliefs that divide the universe into the complementary oppositions of yin and yang. In traditions such as classical Hinduism (Samkhya, Yoga, Vaisheshika and the later Vedanta schools, which accepted the theory of Gunas), Zen Buddhism or Islamic Sufism, a key to enlightenment is "transcending" this sort of dualistic thinking, without merely substituting dualism with monism or pluralism.

The yin and yang symbolizes the duality in nature and all things in the Taoist religion.

Chinese Philosophy

The opposition and combination of the universe's two basic principles of yin and yang is a large part of Chinese philosophy, and is an important feature of Taoism, both as a philosophy and as a religion, although the concept developed much earlier. Some argue that yin and yang were originally an earth and sky god, respectively. As one of the oldest principles in Chinese philosophy, yin and yang are also discussed in Confucianism, but to a lesser extent.

Some of the common associations with yang and yin, respectively, are: male and female, light and dark, active and passive, motion and stillness. Some scholars believe that the two ideas may have originally referred to two opposite sides of a mountain, facing towards and away from the sun. The yin and yang symbol in actuality has very little to do with Western dualism; instead it represents the philosophy of balance, where two opposites co-exist in harmony and are able to transmute into each other. In the yin-yang symbol there is a dot of yin in yang and a dot of yang in yin. In Taoism, this symbolizes the inter-connectedness of the opposite forces as different aspects of Tao, the First Principle. Contrast is needed to create a distinguishable reality, without which we would experience nothingness. Therefore, the independent principles of yin and yang are actually dependent on one another for each other's distinguishable existence.

The complementary dualistic concept seen in yin and yang represent the reciprocal interaction throughout nature, related to a feedback loop, where opposing forces do not exchange in opposition but instead exchange reciprocally to promote stabilization similar to homeostasis. An underlying principle in Taoism states that within every independent entity lies a part of its opposite. Within sickness lies health and vice versa. This is because all opposites are manifestations of the single Tao, and are therefore not independent from one another, but rather a variation of the same unifying force throughout all of nature.

COSMIC INFLATION

Cosmic inflation is a faster-than-light expansion of the universe that spawned many others.

Inflation was invented to explain a couple of features of the universe that are really hard to explain without it. The first is that Einstein's general theory of relativity famously makes mass bend space and time – so you'd expect a universe like ours, containing mass, to be overall curved in some way, either in on itself like a ball ("positive" curvature) or out on itself like a saddle ("negative" curvature).

In fact, it's darn nearly flat. Not only that, but even bits of it far off in different directions as observed from Earth have roughly the same temperature, although in an expanding universe there wouldn't have been time for heat to pass between them to even things out. That seems a naked assault on the laws of thermodynamics.

Cosmic inflation solves these problems at a stroke. In its earliest instants, the universe expanded faster than light (light's speed limit only applies to things within the universe). That ironed out wrinkles in its early chaotic self and meant that even now far-flung parts were once in close contact, so could swap heat.

Inflation is now a built-in piece of our standard story of cosmic evolution. But it's still controversial. In 2014, researchers claimed to have seen ripples from inflation imprinted on the cosmic microwave background. But this proved mistaken, and it's not clear what would have made the early universe inflate anyway. Worse, inflation is very difficult to stop, creating a multiverse of causally disconnected universes that eternally bud off from one another.

One way out might be to weaken the constant speed of light. If the speed of light was faster in the early universe, that would also explain the temperature problem. Perhaps light is still slowing now, just at a rate that is imperceptible even to our most sensitive detectors.

Another conundrum thrown up by the basic Big Bang theory is how to explain the relative homogeneity and evenness of the temperature of the cosmic microwave background radiation. How did large-scale structures such as galaxies and clusters of galaxies develop out of what should have been a rather boringly amorphous and featureless fireball?

This would appear to be in direct violation of the Second Law of Thermodynamics, which describes an inexorable tendency towards entropy and uniformity and away from patterns and structures. If our universe had started out completely smooth, then it should have continued that way, and the universe today would contain nothing more than thinly spread dark matter along with less than one atom per cubic meter of hydrogen and helium gas, with no sign of the texture and complexity we see around us (stars, galaxies, a multitude of elements, life).

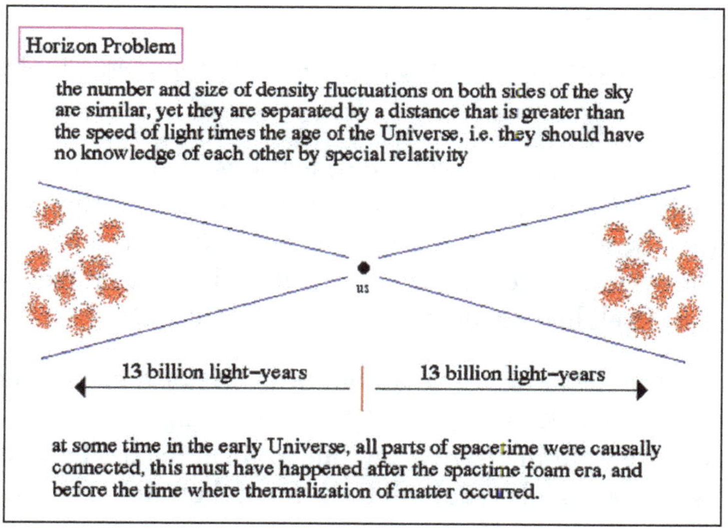

The horizon problem of the Big Bang model.

However, even very slight irregularities in the early phases of expansion would have become amplified as slightly dense patches are affected by additional gravity until they condensed into self-contained structures held together by their own gravity. Galaxies crashed and merged and cannibalized their neighbors, and larger scale structures like clusters and super-clusters formed by a continuing process of gravitational aggregation working on these newly formed galaxies.

Heat tends to travel from a hot body to a cold one so that the temperatures of both bodies eventually even out (a result of the Second Law of Thermodynamics itself), like hot coffee in a cold cup. The microwave background radiation discovered by Arno Penzias and Robert Wilson in the 1960s appeared to be extremely uniform throughout the observable universe, with almost no variance. But if, as the evidence suggests, the last time the cosmic background radiation had any contact with matter was about 450,000 years after the Big Bang (by which time the universe had cooled to around 3,000 °C), then this presents a paradox, because the universe at that time would already have had a diameter of around 90 million light years, and just not enough time would have elapsed for radiation or heat to have flown around the whole universe and equalized itself, and the horizons could never have actually been in causal contact with each other (known as the "horizon problem").

So, in theory, there actually ought to be even more variation today than there is. That is, unless the very early universe was in fact much smaller than the models were predicting. The most widely accepted theory as to how this might have been possible is known as cosmic inflation, which was first proposed in 1980 by the American physicist Alan Guth, developed out of Steven Weinberg's Electroweak Theory and Grand Unified Theory.

The addition of inflation to the Big Bang model claimed to solve the horizon problem, as well as one or two other potential problems that had been identified with the standard Big Bang theory, such as the "flatness problem" (why the density of matter in the universe appears "fine-tuned" to be very close to the critical value at which space is perfectly flat rather than a non-Euclidean hyperbolic or spherical shape) and the "magnetic monopole problem" (why the magnetic monopoles which theory suggests should have been produced in the high temperatures of the early universe appear not to have persisted to the present day).

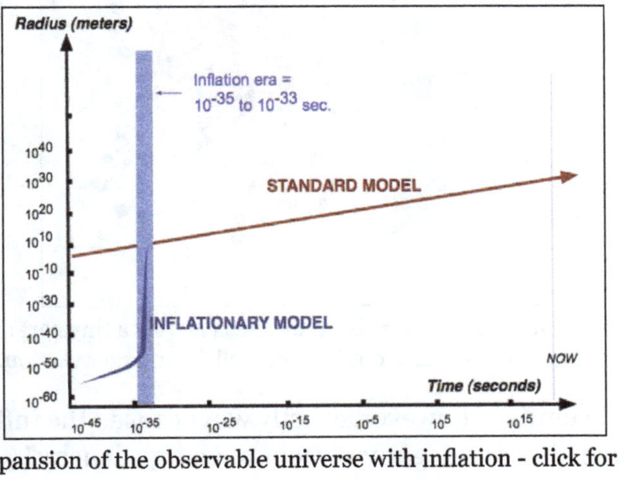

Graph of the expansion of the observable universe with inflation - click for larger version.

Cosmic inflation is the idea that the very early universe went through a period of accelerated, exponential expansion during the first 10-35 of a second before settling down to the more sedate rate of expansion we are still experiencing, so that all of the observable universe originated in a small (indeed, microscopic) causally-connected region.

Although the universe has been expanding since the initial Big Bang, inflation refers to the hypothesis that, for a very short time, the universe expanded at a sharply INCREASING rate, rather than at the decreasing rate it followed before inflation and has followed since. By some calculations, inflation increased the size of the universe by a factor of around 1026 during that tiny fraction (far less than a trillionth) of a second, expanding it from smaller than the size of a proton to about the size of a grapefruit.

Technically, the expansion during this period of inflation (and even the somewhat slower expansion which succeeded it) proceeded faster than the speed of light. To explain how this is possible (the speed of light being supposedly the maximum speed it is possible to travel), an analogy may help. If two airplanes are flying directly away from each other at their maximum speed of, say, 500 kilometers per hour, they are actually flying apart at 1,000 kilometers per hour even though neither individual plane is exceeding 500km per hour. Thus, "expansion", in terms of the expanding universe, is not the same thing as "travel".

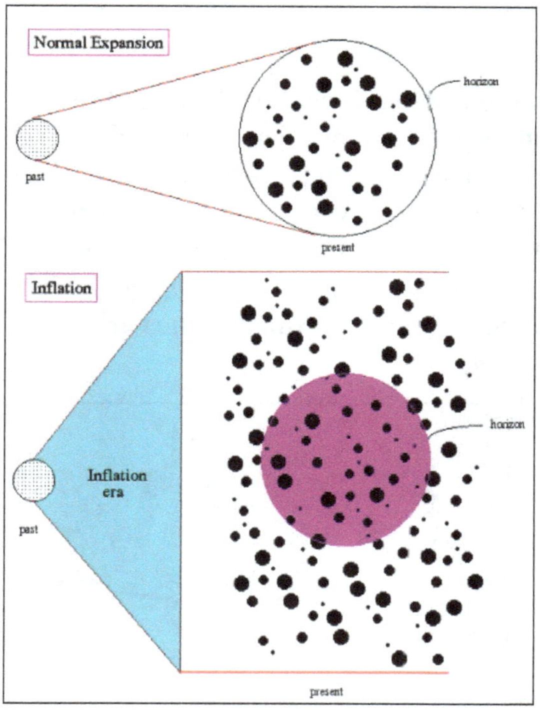

Under inflation, the observable universe is merely a tiny part of the whole that lies within our horizon - click for larger version.

It is still not clear to scientists, however, exactly what caused the inflationary phase, the best guess being some kind of a negative "vacuum energy density" (or positive "vacuum pressure") triggered by the separation of the strong nuclear force from the other elementary forces at this time. It is hypothesized that this separation caused a kind of symmetry breaking or phase transition (analagous to the phase transition when water turns to ice), which left the universe in a highly unstable state with much more energy than it would otherwise have had, causing a sharp outward antigravitational effect, smoothing out most of the irregularities in the existing matter and creating vast quantities of particles in a very short time.

This theory allows for some kind of very slight unevenness (so-called quantum fluctuations) on a sub-atomic scale at a very early stage in the growth of the universe which provided starting points for the large-scale structures we see in today's universe. This suggests the rather bizarre possibility that the largest structures in the universe, the great clusters of galaxies, may actually have been spawned by sub-microscopic seeds.

Guth hypothesized that the reason why the universe appears to be flat is because it is actually fantastically big (in the same way that the spherical Earth appears flat to those on its surface), and that the observable universe is actually only a very small part of the actual universe. In fact, Guth's calculations suggest that the entire universe may be at least 1023 times bigger than the size of the observable universe, roughly equal to the ratio of the size of the observable universe to the planet Earth. Thus, although the observable universe may appear to be effectively flat, the entire universe may be completely different in nature. Also, although an enormous number of magnetic monopoles could well have arisen in the inflationary early universe, the chances of actually observing even one magnetic monopole are infinitesimally small in a universe of such immense size.

Thus, the incredibly vast and fast expansion of the universe caused by inflation "solved" both Robert Dickes flatness problem and Guth's own monopole problem. But it also solved the horizon problem: according to the inflation theory, the universe blew up so quickly that there was no time for the essential homogeneity to be broken, and the universe after inflation would therefore have been very uniform, even though the parts of it were not still in touch with each other.

In an attempt to prove the inflation theory, the Cosmic Background Explorer (COBE) probe was launched in 1992, and its initial results confirmed almost exactly the amount of variation in the cosmic microwave background radiation that was predicted by inflationary theory. In 2003, the Wilkinson Microwave Anisotropy Probe (WMAP) demonstrated the existence of these non-uniformities with even greater precision. As recently as 2014, astronomers at the Harvard-Smithsonian Center for Astrophysics announced that they had detected and mapped "gravitational waves" within the cosmic microwave background radiation, providing further strong evidence for inflation (and for the Big Bang itself), although further peer reivew of these new findings are still ongoing.

Guth's theory has been very influential, even if he himself could find no way to end inflation so that stars and galaxies could form (known the "graceful exit" problem), and he considered his own theory something of a failure because of this. There have been many other refinements and revisions since Guth's original model, such as the "new inflationary model" of Russian physicist Andrei Linde, who had been working on an inflation theory independently (as had Paul Steinhardt and Andreas Albrecht). This new model hypothesized a slow (as opposed to Guth's fast) breaking of symmetry, and the creation of many "bubble universes" (just one of which contains our own observable universe). A later proposal by Linde, known as the "chaotic inflationary model", hypothesized that the repulsive antigravity effect was caused by a "spin-0 field" rather than any kind of phase transition as Guth had thought.

Linde's work, and that of fellow Russian Alex Vilenkin, has also given rise to the idea of "eternal inflation", where the inflation as a whole actually never stops, but small localized energy discharges within the overall energy field - almost like sparks of static electricity, but on on a cosmic scale - create small points of matter in the form of tiny particles. Such a process may represent the birth of a new universe, such as our own. Beginning in this way with what we have called a Big Bang, this new universe then itself proceeds to expand, although at a much slower rate than the continuing inflation outside of it. The rest of space outside of that universe is still full of undischarged energy, still expanding at enormous speed, and new universes, new Big Bangs, are occurring all the time.

The theory of cosmic inflation, then, supports the scenario in which our universe is just one among many parallel universes in a multiverse.

Issues in Cosmology

Age of the Universe

The issue of the age of the Universe was the focus of much discussion during the 19th century, centring on the geological age of the Earth. A major breakthrough was the discovery of the longevity of the Sun, via thermonuclear energy sources, in the first half of the 20th century, coupled with the realization that the expansion age of the Universe exceeded the age of the Sun and was comparable to the ages of the oldest stars in our Galaxy. The scene had shifted: from the age ofthe Earth versus that of the Sun to the ages of the oldest stars versus that of the Big Bang.

During the second halfofthe 20th century, a controversy raged around the value of the Rubble constant, the inverse of which gives a first approximation to the age of the Universe. One school, led by acontroversy raged Gerard de Vaucouleurs, argued vociferously for a high value ofRubaround the value of ble's constant (100 km/Mpc/sec). If de Vaucouleurs and his supporters the Hubble constant had prevailed, the Big Bang would have been in considerable difficulty. However, Allan Sandage, Gustav

Tammann and their supporters were equally strong advocates for a low value of Rubble's constant of 50 km/Mpc/sec. Onlywith the passage oftime and the entry into the fray of a new generation of astronomers who exploited new instruments, most notably the Rubble Space Telescope, did the situation converge. Right precision observations of key distance calibrators such as Cepheid variable stars and supernovae in distant galaxies resulted in a consensus that Rubble's constant is (70 ± 7) km/Mpc/s, a value that poses no problem whatsoever in reconciling the oldest stellar ages with the age of the Universe.

The canonical cosmological model has an age of 14 Gyr. The cosmological age is consistent with two independent age determinations for the Milky Way galaxy. Stellar evolution applied to the oldest globular cluster stars gives an age of 13 Gyr, and radioactive dating from measuring abundances of U and Th in extremely metal-poor halo stars yield an age of 12 Gyr.

There are also cosmological measures of the Rubble constant that are independent of distance. One utilizes the Sunyaev-Zeldovich effect on galaxy clusters to combine X-ray measurements of galaxy clusters with the microwave (Rayleigh Jeans) decrement in the cosmic blackbody background radiation due to traversal across the hot intracluster gas by the microwave photons to give an absolute distance measurement. Yet another approach makes use of the cosmic microwave background radiation anisotropies, where the height of the first acoustic peak is sensitive to the Rubble constant, given a cosmological model.

There is complete consistency between the various Rubble constant determinations. There is no age controversy for the Big Bang.

Galaxy Formation

Galaxy formation theory is not in a very satisfactory state. This stems ultimately from our lack of any fundamental understanding of star formation. There is no robust theory for the detailed properties of galaxies. In contrast, the evolution of the dark matter, which is by far the predominant constituent ofthe Universe, is well understood. Hierarchical formation of large-scale structure in a cold dark matter-dominated Friedmann-Lemaitre Universe has successfully confronted essentially all observations, ranging from deep surveys of the galaxy distribution and the formation of galaxy clusters to the temperature fluctuations in the cosmic microwave background.

The advent of high-resolution N-body simulations has revealed challenges for galaxy formation. Dark halos contain considerable sub-structure, amounting to circa 10% of the halo mass, and continuing down to unresolved scales of 106 M§ or smaller. This has led to two problems. One is that the predicted number of dwarf galaxy satellites exceeds that observed around the Milky Way by an order of magnitude. Even very low surface brightness dwarfs are easily observed in our local environment, and efficient mass loss prior to star formation has been invoked to resolve this discrepancy.

A second problem is that, on the one hand, the initial angular momentum of a typical protogalaxy, when the halo first collapsed, can reproduce the observed size distribution of disks, if angular momentum is approximately conserved during baryon infall and disk accretion, but that, on the other hand, the numerical simulations show' that most of the angular momentum is actually lost to the dark halo. The clumpiness induces strong angular momentum transfer via tidal torqueing and dynamical friction from the dissipating baryons to the energy-conserving dark matter. The resulting disks have too little specific angular momentum by an order of magnitude.

Another result from the high-resolution simulations of galaxy halos is that the dark halo has a central concentration and a central cusp. This implies for example that about half the mass within the half-light radius (for our Galaxy this corresponds to near the solar circle) is in the form of cold dark matter (CDM). This apparently contradicts bulge microlensing studies for the Milky Way, which permit a CDM fraction of at most 10% within the solar circle, when combined with the rotation curve and with infrared stellar population modelling derived from the results of the Deep Infrared Background Experiment (DIRBE) that was flown on the Cosmic Background Explorer (COBE). A similar conclusion applies to barred galaxies, for which dynamical studies conclude that self-gravity of the bars, which consist primarily of stars, must dominate the inner gravitational potential. In contrast, a recent study of a non-barred spiral finds that detailed modelling of the rotation curve with a maximal disk requires an axially-symmetric dark-matter contribution of about 30% within an optical radius.

There is another observational hurdle to overcome. Low surface brightness spirals are everywhere dark-matter dominated, and so provide outstanding laboratories for dark matter studies via rotation curves. High-resolution Ra rotation curves reveal a wide array of central profiles. Most systems have soft cores without any indication of a central cusp.

Cold dark matter is seriously challenged. Whether it is actually dead is quite another matter. Nevertheless, to confront this possibility, there have been numerous attempts to resurrect CDM. These come under two distinct guises: tinkering with the particle physics or elaborating on the astrophysics. Modified-particle dark-matter can no doubt be developed to explain all of the required dark matter properties. However the seductive simplicity of SUSY, with the lightest stable supersymmetric particle, the neutralino, as an attractive candidate for CDM, is lost.

An alternative approach is via the astrophysics of galaxy formation. Can the dark matter profile be modified by astrophysical processes? The answer is: perhaps.

Supermassive black holes, for better or for worse, are intimately connected with the process of galaxy spheroid formation. Whether they aid and abet formation of the first stars remains a mystery. There are certainly dynamical and most likely astrochemical links. The supermassive black-hole (SMBH) merger results in heating of the dark matter cusp. The dark matter is heated and acquires angular momentum. The result is that the concentration and cusp are likely to be modified. However the angular momentum acquired by the halo will help in reinjecting angular momentum into infalling gas clouds that form the disk over a time-scale of a Gyr or longer.

It remains to be seen whether the ultimate answer lies in the dynamical feedback of 5MBH formation and evolution on dark halo cores, or on a new prescription that modifies the physics of CDM, or possibly in fundamental physics whereby on large scales unanticipated changes in our 4-d Einstein gravity may be appearing, such as might be associated with the influence of higher dimensions.

Dark Matter

The favoured candidate for CDM is the lightest stable SUSY relic particle. This must be neutral (to avoid already having been detected) and its mass is constrained by accelerator searches and theoretical considerations of thermal freeze-out to lie in the range 50 GeV to a few TeV. The relic density is determined when annihilations and pair production go out of thermal equilibrium in the early Universe at $T \approx m_x / 20kK$, and one infers that the density in relic particles, relative to the critical density of an Einstein- de Sitter Universe, satisfies $\Omega_x \propto \sigma_{ann}$, where σ_{ann} is the annihilation cross-section extrapolated to the low temperature limit and m_x is the mass of the relic particle. For typical weak interaction values of σ_{ann}, one finds that $\Omega_x \approx 0.3$ is required to account for the dark matter content of the Universe. Via studying a grid of supersyrnrnetric (SUSY) models, one can infer a range of particle masses from the annihilation cross-section. Were it not for the accelerator bounds on the sparticle (i.e., supersyrnmetric partner-particle) masses, the uncertainty in m_x would span some 5 orders of magnitude.

The annihilation cross-section and particle mass is constrained. So also is the elastic scattering cross-section once the annihilation cross-section is specified. This means that one can now consider possible detection schemes. The obvious one is direct detection by elastic scattering. Use of annual modulation of the incident flux on a terrestrial detector has led to a tentative detection (DAMA)[2] that requires an implausibly large cross-section given the suite of minimal SUSY models and is marginally inconsistent with another experiment (CDMS)[3]. Annihilations result in hadronic jets that decay into gamma rays, high-energy electron-positron pairs, proton-antiproton pairs and neutrinos, all of which are potentially detectable as galactic halo signals.

Future observations may greatly help in pinning down the CDM characteristics. Annihilations generate high-energy neutrinos and gamma rays that propagate freely in the halo and detection would provide unambiguous support for annihilating Weakly Interacting Massive Particles (WIMPs). The predicted fluxes are within the anticipated sensitivity of the ANTARES, AMANDA (neutrino) and GLAST (gamma ray) detectors, now under construction.

References

- Cosmology, cosmos: astronomy.swin.edu.au, Retrieved 13 June, 2020
- Relativistic-cosmologies, science-cosmology-astronomy: britannica.com, Retrieved 28 April, 2020
- Cosmic-inflation, term: newscientist.com, Retrieved 12 May, 2020
- Topics-bigbang-inflation: physicsoftheuniverse.com, Retrieved 05 January, 2020
- Lavazza, Andrea; Robinson, Howard, eds. (2014). Contemporary Dualism: A Defense. Routledge Studies in Contemporary Philosophy. New York: Routledge. ISBN 978-0-415-81882-7

3
Physical Cosmology

The branch of cosmology that deals with the study of largest-scale dynamics and structure of the universe is referred to as physical cosmology. Dark matter, plasma cosmology, observational cosmology, Copernican principle, inhomogeneous cosmology, etc. fall under its domain. This chapter discusses physical cosmology in detail.

Physical cosmology, as a branch of astrophysics, is the study of the large-scale structure of the universe and is concerned with fundamental questions about its formation and evolution.

Cosmology involves itself with studying the motions of the celestial bodies and the first cause.

Physical cosmology, as it is now understood, began with the twentieth century development of Albert Einstein's theory of general relativity and detailed astronomical observations of extremely distant objects.

Cosmology has developed into a science characterised by precision through ever more accurate measurements of temperature anisotropies in the Cosmic Microwave Background (CMB), along with studies of the expansion history of the Universe, as well as sky surveys providing detailed mapping of large-scale structures.

This exciting development has been possible thanks to ground-breaking discoveries in the theoretical framework of cosmology over the past half century. This year's Nobel Laureate James Peebles has made seminal contributions in this area. Through detailed modelling, with the help of analytic as well as numerical methods, he has explored fundamental properties of our Universe and uncovered unexpected new physics. We now have a unified model capable of describing the Universe from its earliest fraction of a second up until the present and into the distant future.

Modern cosmology is based on Einstein's theory of general relativity and assumes an early era, the Big Bang, when the Universe was extremely hot and dense. A little less than 400,000 years after the Big Bang, the temperature had decreased to about 3,000 K, enabling electrons to combine with nuclei into atoms. Because no charged particles were left that could easily interact with the photons, the Universe became transparent to light. This radiation is now visible as the CMB. Due to the cosmological redshift, its temperature is currently just 2.7 K — a factor of about 1,100 lower since the decoupling of matter and radiation. In figure 1, the source of the CMB can be seen as a screen that prevents us from easily looking back in time further than to a few hundred thousand years after the Big Bang.

A timeline of our Universe extending from an unknown origin on the left to a darkening future on the right.

One of the very first to propose that the Universe started with something like a Big Bang was the American horror writer Edgar Allan Poe in his prose poem Eureka". As an explanation as to why the sky is dark at night, often referred to as Olber's paradox after the German astronomer Heinrich Wilhelm Olbers, Poe suggested that the Universe had a beginning. The first to formulate a mathematical theory for the expanding Universe, using Einstein's newly developed theory of general relativity, was the Russian mathematician and cosmologist Alexander Friedman in 1922. He further developed his theory in 1924. These ideas were rediscovered in 1927 by the Belgian Catholic priest and astronomer Georges Lemaître, who later introduced the notion of a "primeval atom". He argued that the galaxies were receding from each other, and that this could be explained if the Universe expanded. In 1924, the Swedish astronomer Knut Lundmark had made a similar observation, albeit with less rigor and accuracy. A more general acceptance that the Universe was in fact expanding came with the observations by the US astronomer Edwin Hubble in 1929.

It is easy to derive the basic equations that describe the expansion of the Universe, the Friedman equations, even without the use of general relativity. To see this, let us for simplicity assume a homogenous universe. We pick an arbitrary point, at rest relative to matter, draw a sphere around it with radius R, and assume the sphere will grow as the universe expands. On the surface of the sphere, we introduce a small test mass with mass m. The total energy of the test mass is given by:

$$E = \frac{m\dot{R}^2}{2} - \frac{GMm}{R}$$

where $M = \frac{4\pi}{3}\rho R^3$. A simple rearrangement gives:

$$\left(\frac{\dot{R}}{R}\right)^2 = \frac{8\pi G}{3}\rho - \frac{kc^2}{R^2}$$

where $k = -\frac{2E}{mc^2}$. Identifying $H = \frac{\dot{R}}{R}$ as the Hubble constant, this becomes the first Friedman equation. By rescaling R one can set $k = \pm 1, 0$. To correctly interpret the meaning of k, we need to appeal to general relativity, where it is identified as the spatial curvature. The value $k=0$ corresponds to the critical density of a flat universe given by:

$$\rho_c = \frac{3H^2}{8\pi G}.$$

Observations show that the total energy density of the Universe is very close to this value. Defining $\Omega = \frac{\rho}{\rho_c}$, we have $\Omega < 1$ for a universe with negative curvature, $\Omega = 1$ for a flat universe, and $\Omega > 1$ for a universe with positive curvature.

There are several different components of energy in the Universe. Matter in the form of pressure less dust has an energy density that dilutes with volume, described by $1/R^3$, while radiation disperses according to $1/R^4$, due to the loss of energy caused by redshift. In the early Universe, radiation dominated the energy density of the Universe until a bit before recombination. Moreover, in the framework of general relativity, and to account for the possibility that the Universe could have been static, Einstein introduced an additional term in 1917, corresponding to a constant energy contributing to ρ, the cosmological constant, Λ.

Multiplying the Friedman equation with R^2, to think of it as energy conservation, makes it easy to figure out what is actually happening. We see the effective gravitational potential in the case of matter or radiation. When $k > 0$, the Universe reaches a maximum size and then re-contracts. If $k < 0$, it may keep on expanding forever.

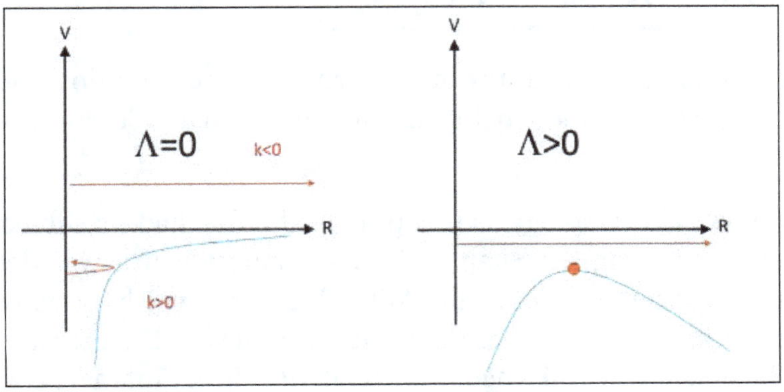

The effective gravitational potential V as a function of the scale factor R, without a cosmological constant on the left and with a cosmological constant contributing to the energy density on the right. The red arrows show what happens in a few cases. The red dot represents Einstein's static, and unstable, universe.

From observations, we know that the amount of ordinary baryonic matter in the form of nucleons, present in stars, clouds of gas, and the like, is no more than 5% of the total energy density at present. In addition, dark matter contributes 26% of the Universe's critical density. It might as well be called invisible matter, as it neither emits nor absorbs light and is so far only known through its gravitational effects.

The most important component is the cosmological constant, representing a constant energy density unaffected by expansion. It is often called dark energy to account for the possibility that it might vary over time and space. That is, dark energy is not necessarily the constant introduced and thought to be related to the vacuum energy in quantum field theory. Observations show that dark energy contributes the remaining 69% of the critical density. As the other components of matter become diluted by the expansion, the dark energy will become ever more important with time (unless its energy density starts to decrease).

A cosmological constant has been added to the potential with dramatic results. We note how the potential slopes downwards and gives rise to the accelerated expansion. The static universe of Einstein is marked as the unstable point at the maximum of the potential. That dark energy can force galaxies to accelerate away from each other may seem counterintuitive, but it is a direct consequence of the unusual properties of dark energy. The accelerated phase, where the number of galaxies thins out.

The hot Big Bang is the fire in the middle of the diagram, indicating that a preparatory phase, such as inflation, might exist before the hot Big Bang. Inflation is postulated as a scenario with a period of rapid acceleration, which would explain several properties of our Universe, such as its flatness.

Observational Cosmology

Observational cosmology is the study of the structure, the evolution and the origin of the universe through observation, using instruments such as telescopes and cosmic ray detectors.

The science of physical cosmology as it is practiced today had its subject material defined in the years following the Shapley-Curtis debate when it was determined that the universe had a larger scale than the Milky Way galaxy. This was precipitated by observations that established the size and the dynamics of the cosmos that could be explained by Einstein's General Theory of Relativity. In its infancy, cosmology was a speculative science based on a very limited number of observations and characterized by a dispute between steady state theorists and promoters of Big Bang cosmology. It was not until the 1990s and beyond that the astronomical observations would be able to eliminate competing theories and drive the science to the "Golden Age of Cosmology" which was heralded by David Schramm at a National Academy of Sciences colloquium in 1992.

Hubble's Law and the Cosmic Distance Ladder

Distance measurements in astronomy have historically been and continue to be confounded by considerable measurement uncertainty. In particular, while stellar parallax can be used to measure the distance to nearby stars, the observational limits imposed by the difficulty in measuring the minuscule parallaxes associated with objects beyond our galaxy meant that astronomers had to look for alternative ways to measure cosmic distances. To this end, a standard candle measurement for Cepheid variables was discovered by Henrietta Swan Leavitt in 1908 which would provide Edwin Hubble with the rung on the cosmic distance ladder he would need to determine the distance to spiral nebula. Hubble used the 100-inch Hooker Telescope at Mount Wilson Observatory to identify individual stars in those galaxies, and determine the distance to the galaxies by isolating individual Cepheids. This firmly established the spiral nebula as being objects well outside the Milky Way galaxy. Determining the distance to "island universes", as they were dubbed in the popular media, established the scale of the universe and settled the Shapley-Curtis debate once and for all.

Astronomer Edwin Hubble.

In 1927, by combining various measurements, including Hubble's distance measurements and Vesto Slipher's determinations of redshifts for these objects, Georges Lemaître was the first to estimate a constant of proportionality between galaxies' distances and what was termed their "recessional velocities", finding a value of about 600 km/s/Mpc. He showed that this was theoretically expected in a universe model based on general relativity. Two years later, Hubble showed that the relation between the distances and velocities was a positive correlation and had a slope of about 500 km/s/Mpc. This correlation would come to be known as *Hubble's law* and would serve as the observational foundation for the expanding universe theories on which cosmology is still based. The publication of the observations by Slipher, Wirtz, Hubble and their colleagues and the acceptance by the theorists of their theoretical implications in light of Einstein's General theory of relativity is considered the beginning of the modern science of cosmology.

Nuclide Abundances

Determination of the cosmic abundance of elements has a history dating back to early spectroscopic measurements of light from astronomical objects and the identification of emission and absorption lines which corresponded to particular electronic transitions in chemical elements identified on Earth. For example, the element Helium was first identified through its spectroscopic signature in the Sun before it was isolated as a gas on Earth.

Computing relative abundances was achieved through corresponding spectroscopic observations to measurements of the elemental composition of meteorites.

Detection of the Cosmic Microwave Background

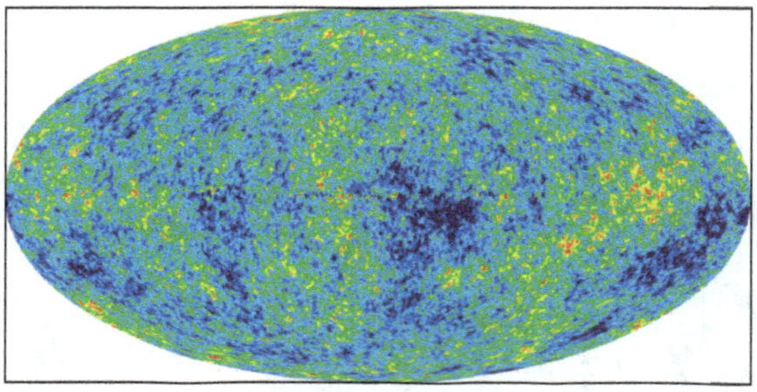

The CMB seen by WMAP.

A cosmic microwave background was predicted in 1948 by George Gamow and Ralph Alpher, and by Alpher and Robert Herman as due to the hot Big Bang model. Moreover, Alpher and Herman were able to estimate the temperature, but their results were not widely discussed in the community. Their prediction was rediscovered by Robert Dicke and Yakov Zel'dovich in the early 1960s with the first published recognition of the CMB radiation as a detectable phenomenon appeared in a brief paper by Soviet astrophysicists A. G. Doroshkevich and Igor Novikov, in the spring of 1964. In 1964, David Todd Wilkinson and Peter Roll, Dicke's colleagues at Princeton University, began constructing a Dicke radiometer to measure the cosmic microwave background. In 1965, Arno Penzias and Robert Woodrow Wilson at the Crawford Hill location of Bell Telephone Laboratories in nearby Holmdel Township, New Jersey had built a Dicke radiometer that they intended to use for radio astronomy and satellite communication experiments. Their instrument had an excess 3.5 K antenna temperature which they could not account for. After receiving a telephone call from Crawford Hill, Dicke famously quipped: "Boys, we've been scooped." A meeting between the Princeton and Crawford Hill groups determined that the antenna temperature was indeed due to the microwave background. Penzias and Wilson received the 1978 Nobel Prize in Physics for their discovery.

Modern Observations

Today, observational cosmology continues to test the predictions of theoretical cosmology and has led to the refinement of cosmological models. For example, the observational evidence for dark matter has heavily influenced theoretical modeling of structure and galaxy formation. When trying to calibrate the Hubble diagram with accurate supernova standard candles, observational evidence for dark energy was obtained in the late 1990s. These observations have been incorporated into a six-parameter framework known as the Lambda-CDM model which explains the evolution of the universe in terms of its constituent material. This model has subsequently been verified by detailed observations of the cosmic microwave background, especially through the WMAP experiment.

Included here are the modern observational efforts that have directly influenced cosmology.

Redshift Surveys

With the advent of automated telescopes and improvements in spectroscopes, a number of collaborations have been made to map the universe in redshift space. By combining redshift with angular position data, a redshift survey maps the 3D distribution of matter within a field of the sky. These observations are used to measure properties of the large-scale structure of the universe. The Great Wall, a vast supercluster of galaxies over 500 million light-years wide, provides a dramatic example of a large-scale structure that redshift surveys can detect.

The first redshift survey was the CfA Redshift Survey, started in 1977 with the initial data collection completed in 1982. More recently, the 2dF Galaxy Redshift Survey determined the large-scale structure of one section of the Universe, measuring z-values for over 220,000 galaxies; data collection was completed in 2002, and the final data set was released 30 June 2003. (In addition to mapping large-scale patterns of galaxies, 2dF established an upper limit on neutrino mass.) Another notable investigation, the Sloan Digital Sky Survey (SDSS), is ongoing as of 2011 and aims to obtain measurements on around 100 million objects. SDSS has recorded redshifts for galaxies as high as 0.4, and has been involved in the detection of quasars beyond $z = 6$. The DEEP2 Redshift Survey uses the Keck telescopes with the new "DEIMOS" spectrograph; a follow-up to the pilot program DEEP1, DEEP2 is designed to measure faint galaxies with redshifts 0.7 and above, and it is therefore planned to provide a complement to SDSS and 2dF.

Cosmic Microwave Background Experiments

Subsequent to the discovery of the CMB, hundreds of cosmic microwave background experiments had been conducted to measure and characterize the signatures of the radiation. The most famous experiment is probably the NASA Cosmic Background Explorer (COBE) satellite that orbited in 1989–1996 and which detected and quantified the large-scale anisotropies at the limit of its detection capabilities. Inspired by the initial COBE results of an extremely isotropic and homogeneous background, a series of ground-based and balloon-based experiments quantified CMB anisotropies on smaller angular scales over the next decade. The primary goal of those experiments was to measure the angular scale of the first acoustic peak, for which COBE did not have sufficient resolution. The measurements were able to rule out cosmic strings as the leading theory of cosmic structure formation, and suggested cosmic inflation was the right theory. During the 1990s, the first peak was measured with increasing sensitivity and by 2000 the BOOMERanG experiment reported that the highest power fluctuations occur at scales of approximately one degree. Together with other cosmological data, these results implied that the geometry of the Universe is flat. A number of ground-based interferometers provided measurements of the fluctuations with higher accuracy over the next three years, including the Very Small Array, Degree Angular Scale Interferometer (DASI) and the Cosmic Background Imager (CBI). DASI made the first detection of the polarization of the CMB and the CBI provided the first E-mode spectrum with compelling evidence that it is out of phase with the T-mode spectrum.

In June 2001, NASA launched a second CMB space mission, WMAP, to make much more precise measurements of the large-scale anisotropies over the full sky. The first results from this mission, disclosed in 2003, were detailed measurements of the angular power spectrum to below degree scales, tightly constraining various cosmological parameters. The results are broadly consistent with those expected from cosmic inflation as well as various other competing theories, and are available in detail at NASA's data center for Cosmic Microwave Background (CMB). Although WMAP provided very accurate measurements of the large angular-scale fluctuations in the CMB (structures about as large in the sky as the Moon), it did not have the angular resolution to measure the smaller scale fluctuations which had been observed using previous ground-based interferometers.

A third space mission, Planck, was launched in May 2009. Planck employs both HEMT radiometers and bolometer technology and measures the CMB anisotropies at a higher resolution than WMAP. Unlike the previous two space missions, Planck is a collaboration between NASA and the European Space Agency (ESA). Its detectors got a trial run at the Antarctic Viper telescope as ACBAR (Arcminute Cosmology Bolometer Array Receiver) experiment – which has produced the most precise measurements at small angular scales to date – and at the Archeops balloon telescope.

Additional ground-based instruments such as the South Pole Telescope in Antarctica and the proposed Clover Project, Atacama Cosmology Telescope and the QUIET telescope in Chile will provide additional data not available from satellite observations, possibly including the B-mode polarization.

Telescope Observations

Radio

The brightest sources of low-frequency radio emission (10 MHz and 100 GHz) are radio galaxies which can be observed out to extremely high redshifts. These are subsets of the active galaxies that have extended features known as lobes and jets which extend away from the galactic nucleus distances on the order of megaparsecs. Because radio galaxies are so bright, astronomers have used them to probe extreme distances and early times in the evolution of the universe.

Infrared

Far infrared observations including submillimeter astronomy have revealed a number of sources at cosmological distances. With the exception of a few atmospheric windows, most of infrared light is blocked by the atmosphere, so the observations generally take place from balloon or space-based instruments. Current observational experiments in the infrared include NICMOS, the Cosmic Origins Spectrograph, the Spitzer Space Telescope, the Keck Interferometer, the Stratospheric Observatory For Infrared Astronomy, and the Herschel Space Observatory. The next large space telescope planned by NASA, the James Webb Space Telescope will also explore in the infrared.

An additional infrared survey, the Two-Micron All Sky Survey, has also been very useful in revealing the distribution of galaxies, similar to other optical surveys described below.

Optical Rays (Visible to Human Eyes)

Optical light is still the primary means by which astronomy occurs, and in the context of cosmology, this means observing distant galaxies and galaxy clusters in order to learn about the large scale structure of the Universe as well as galaxy evolution. Redshift surveys have been a common means by which this has been accomplished with some of the most famous including the 2dF Galaxy Redshift Survey, the Sloan Digital Sky Survey, and the upcoming Large Synoptic Survey Telescope. These optical observations generally use either photometry or spectroscopy to measure the redshift of a galaxy and then, via Hubble's Law, determine its distance modulo redshift distortions due to peculiar velocities. Additionally, the position of the galaxies as seen on the sky in celestial coordinates can be used to gain information about the other two spatial dimensions.

Very deep observations (which is to say sensitive to dim sources) are also useful tools in cosmology. The Hubble Deep Field, Hubble Ultra Deep Field, Hubble Extreme Deep Field, and Hubble Deep Field South are all examples of this.

Copernican Principle

In physical cosmology, the Copernican principle states that humans, on the Earth or in the Solar System, are not privileged observers of the universe.

Named for Copernican heliocentrism, it is a working assumption that arises from a modified cosmological extension of Copernicus's argument of a moving Earth. In some sense, it is equivalent to the mediocrity principle.

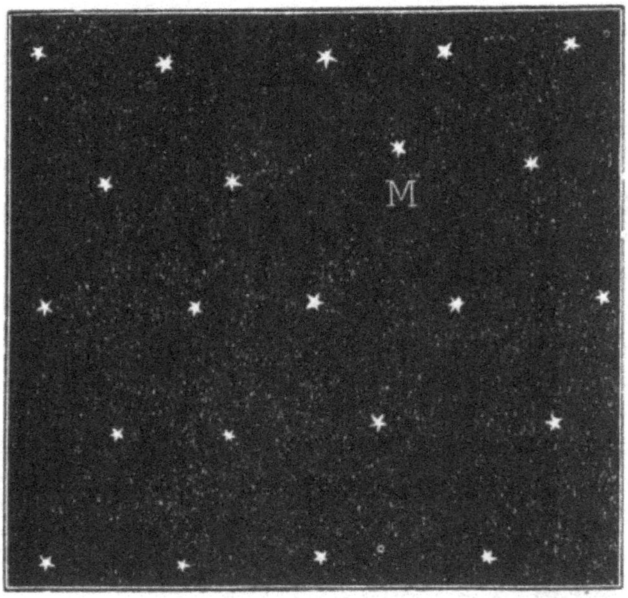

Hermann Bondi named the principle after Copernicus in the mid-20th century, although the principle itself dates back to the 16th-17th century paradigm shift away from the Ptolemaic system, which placed Earth at the center of the universe. Copernicus proposed that the motion of the planets could be explained by reference to an assumption that the Sun is centrally located and stationary on contrast to the then currently upheld belief that the Earth was central. He argued that the apparent retrograde motion of the planets is an illusion caused by Earth's movement around the Sun, which the Copernican model placed at the centre of the universe. Copernicus himself was mainly motivated by technical dissatisfaction with the earlier system and not by support for any mediocrity principle. In fact, although the Copernican heliocentric model is often described as "demoting" Earth from its central role it had in the Ptolemaic geocentric model, it was successors to Copernicus, notably the 16th century Giordano Bruno, who adopted this new perspective. The Earth's central position had been interpreted as being in the "lowest and filthiest parts". Instead, as Galileo said, the Earth is part of the "dance of the stars" rather than the "sump where the universe's filth and ephemera collect". In the late 20th Century, Carl Sagan asked, "Who are we? We find that we live on an insignificant planet of a humdrum star lost in a galaxy tucked away in some forgotten corner of a universe in which there are far more galaxies than people."

In cosmology, if one assumes the Copernican principle and observes that the universe appears isotropic or the same in all directions from the vantage point of Earth, then one can infer that the universe is generally homogeneous or the same everywhere (at any given time) and is also isotropic about any given point. These two conditions make up the cosmological principle. In practice, astronomers observe that the universe has heterogeneous or non-uniform structures up to the scale of galactic superclusters, filaments and great voids. It becomes more and more homogeneous and isotropic when observed on larger and larger scales, with little detectable structure on scales of more than about 200 million parsecs. However, on scales comparable to the radius of the observable universe, we see systematic changes with distance from Earth. For instance, galaxies contain more young stars and are less clustered, and quasars appear more numerous. While this might suggest that Earth is at the center of the universe, the Copernican principle requires us to interpret it as evidence for the evolution of the universe with time: this distant light has taken most of the age of the universe to reach Earth and shows the universe when it was young. The most distant light of all, cosmic microwave background radiation, is isotropic to at least one part in a thousand.

Modern mathematical cosmology is based on the assumption that the Cosmological principle is almost, but not exactly, true on the largest scales. The Copernican principle represents the irreducible philosophical assumption needed to justify this, when combined with the observations.

Michael Rowan-Robinson emphasizes the Copernican principle as the threshold test for modern thought, asserting that: "It is evident that in the post-Copernican era of human history, no well-informed and rational person can imagine that the Earth occupies a unique position in the universe."

Bondi and Thomas Gold used the Copernican principle to argue for the perfect cosmological principle which maintains that the universe is also homogeneous in time, and is the basis for the steady-state cosmology. However, this strongly conflicts with the evidence for cosmological evolution, the universe has progressed from extremely different conditions at the Big Bang, and will continue to progress toward extremely different conditions, particularly under the rising influence of dark energy, apparently toward the Big Freeze or Big Rip.

Since the 1990s the term has been used (interchangeably with "the Copernicus method") for J. Richard Gott's Bayesian-inference-based prediction of duration of ongoing events, a generalized version of the Doomsday argument.

Tests of the Principle

The Copernican principle has never been proven, and in the most general sense cannot be proven, but it is implicit in many modern theories of physics. Cosmological models are often derived with reference to the Cosmological principle, slightly more general than the Copernican principle, and many tests of these models can be considered tests of the Copernican principle.

Before the term Copernican principle was even coined, Earth was repeatedly shown not to have any special location in the universe. The Copernican Revolution dethroned Earth to just one of many planets orbiting the Sun. Proper motion was mentioned by Halley. William Herschel found that the Solar System is moving through space within our disk-shaped Milky Way galaxy. Edwin Hubble showed that the Milky Way galaxy is just one of many galaxies in the universe. Examination of the galaxy's position and motion in the universe led to the Big Bang theory and the whole of modern cosmology.

Modern Tests

Recent and planned tests relevant to the cosmological and Copernican principles include:

- Time drift of cosmological redshifts.
- Modeling the local gravitational potential using reflection of cosmic microwave background (CMB) photons.
- The redshift dependence of the luminosity of supernovae.
- The kinetic Sunyaev–Zel'dovich effect in relation to dark energy.
- Cosmic neutrino background.
- The integrated Sachs–Wolfe effect.
- Testing the isotropy and homogeneity of the CMB.

Physics without the Principle

The standard model of cosmology, the Lambda-CDM model, assumes the Copernican principle and the more general Cosmological principle. The Lambda-CDM model's observations are largely consistent, but there remain unsolved problems. Some cosmologists and theoretical physicists have created models without the Cosmological or Copernican principles to constrain the values of observational results, to address specific known issues, and to propose tests to distinguish between current models and other possible models.

A prominent example in this context is the observed accelerating universe and cosmological constant. Instead of using the current accepted idea of dark energy, this model proposes the universe is much more inhomogeneous than currently assumed, and instead, we are in an extremely large low-density void. To match observations we would have to be very close to the centre of this void, immediately contradicting the Copernican principle.

Non-standard Cosmology

A non-standard cosmology is any physical cosmological model of the universe that was, or still is, proposed as an alternative to the then-current standard model of cosmology. The term non-standard is applied to any theory that does not conform to the scientific consensus. Because the term depends on the prevailing consensus, the meaning of the term changes over time. For example, hot dark matter would not have been considered non-standard in 1990, but would be in 2010. Conversely, a non-zero cosmological constant resulting in an accelerating universe would have been considered non-standard in 1990, but is part of the standard cosmology in 2010.

Several major cosmological disputes have occurred throughout the history of cosmology. One of the earliest was the Copernican Revolution, which established the heliocentric model of the Solar System. More recent was the Great Debate of 1920, in the aftermath of which the Milky Way's status as but one of the Universe's many galaxies was established. From the 1940s to the 1960s, the astrophysical community was equally divided between supporters of the Big Bang theory and supporters of a rival steady state universe; this was eventually decided in favour of the Big Bang theory by advances in observational cosmology in the late 1960s. The current standard model of cosmology is the Lambda-CDM model, wherein the Universe is governed by General Relativity, began with a Big Bang and today is a nearly-flat universe that consists of approximately 5% baryons, 27% cold dark matter, and 68% dark energy.

Lambda-CDM has been an extremely successful model, but retains some weaknesses (such as the dwarf galaxy problem). Research on extensions or modifications to Lambda-CDM, as well as fundamentally different models, is ongoing. Topics investigated include quintessence, Modified Newtonian Dynamics (MOND) and its relativistic generalization TeVeS, and warm dark matter.

Modern physical cosmology as it is currently studied first emerged as a scientific discipline in the period after the Shapley–Curtis debate and discoveries by Edwin Hubble of a cosmic distance ladder when astronomers and physicists had to come to terms with a universe that was of a much larger scale than the previously assumed galactic size. Theorists who successfully developed cosmologies applicable to the larger-scale universe are remembered today as the founders of modern cosmology. Among these scientists are Arthur Milne, Willem de Sitter, Alexander Friedman, Georges Lemaître, and Albert Einstein himself.

After confirmation of the Hubble's law by observation, the two most popular cosmological theories became the Steady State theory of Hoyle, Gold and Bondi, and the big bang theory of Ralph Alpher, George Gamow, and Robert Dicke with a small number of supporters of a smattering of alternatives. Since the discovery of the Cosmic microwave background radiation (CMB) by Arno Penzias and Robert Wilson in 1965, most cosmologists concluded that observations were best explained by the big bang model. Steady State theorists and other non-standard cosmologies were then tasked with providing an explanation for the phenomenon if they were to remain plausible. This led to original approaches including integrated starlight and cosmic iron whiskers, which were meant to provide a source for a pervasive, all-sky microwave background that was not due to an early universe phase transition.

Artist depiction of the WMAP spacecraft at the L2 point. Data gathered by this spacecraft has been successfully used to parametrize the features of standard cosmology, but complete analysis of the data in the context of any non-standard cosmology has not yet been achieved.

Scepticism about the non-standard cosmologies' ability to explain the CMB caused interest in the subject to wane since then, however, there have been two periods in which interest in non-standard cosmology has increased due to observational data which posed difficulties for the big bang. The first occurred was the late 1970s when there were a number of unsolved problems, such as the horizon problem, the flatness problem, and the lack of magnetic monopoles, which challenged the big bang model. These issues were eventually resolved by cosmic inflation in the 1980s. This idea subsequently became part of the understanding of the big bang, although alternatives have been proposed from time to time. The second occurred in the mid-1990s when observations of the ages of globular clusters and the primordial helium abundance, apparently disagreed with the big bang. However, by the late 1990s, most astronomers had concluded that these observations did not challenge the big bang and additional data from COBE and the WMAP, provided detailed quantitative measures which were consistent with standard cosmology.

In the 1990s, a dawning of a "golden age of cosmology" was accompanied by a startling discovery that the expansion of the universe was, in fact, accelerating. Previous to this, it had been assumed that matter either in its visible or invisible dark matter form was the dominant energy density in the universe. This "classical" big bang cosmology was overthrown when it was discovered that nearly 70% of the energy in the universe was attributable to the cosmological constant, often referred to as "dark energy". This has led to the development of a so-called concordance ΛCDM model which combines detailed data obtained with new telescopes and techniques in observational astrophysics with an expanding, density-changing universe. Today, it is more common to find in the scientific literature proposals for "non-standard cosmologies" that actually accept the basic tenets of the big bang cosmology, while modifying parts of the concordance model. Such theories include alternative models of dark energy, such as quintessence, phantom energy and some ideas in brane cosmology; alternative models of dark matter, such as modified Newtonian dynamics; alternatives or extensions to inflation such as chaotic inflation and the ekpyrotic model; and proposals to supplement the universe with a first cause, such as the Hartle–Hawking boundary condition, the cyclic model, and the string landscape. There is no consensus about these ideas amongst cosmologists, but they are nonetheless active fields of academic inquiry.

Today, heterodox non-standard cosmologies are generally considered unworthy of consideration by cosmologists while many of the historically significant nonstandard cosmologies are considered to have been falsified. The essentials of the big bang theory have been confirmed by a wide range of complementary and detailed observations, and no non-standard cosmologies have reproduced the range of successes of the big bang model. Speculations about alternatives are not normally part of research or pedagogical discussions, except as object lessons or for their historical importance. An open letter started by some remaining advocates of non-standard cosmology has affirmed that: "today, virtually all financial and experimental resources in cosmology are devoted to big bang studies."

Alternative Gravity

General relativity, upon which the FRW metric is based, is an extremely successful theory which has met every observational test so far. However, at a fundamental level it is incompatible with quantum mechanics, and by predicting singularities, it also predicts its own breakdown. Any alternative theory of gravity would imply immediately an alternative cosmological theory since current modeling is dependent on general relativity as a framework assumption. There are many different motivations to modify general relativity, such as to eliminate the need for dark matter or dark energy, or to avoid such paradoxes as the firewall.

Machian Universe

Ernst Mach developed a kind of extension to general relativity which proposed that inertia was due to gravitational effects of the mass distribution of the universe. This led naturally to speculation about the cosmological implications for such a proposal. Carl Brans and Robert Dicke were able to successfully incorporate Mach's principle into general relativity which admitted for cosmological solutions that would imply a variable mass. The homogeneously distributed mass of the universe would result in a roughly scalar field that permeated the universe and would serve as a source for Newton's gravitational constant; creating a theory of quantum gravity.

MOND

Modified Newtonian Dynamics (MOND) is a relatively modern proposal to explain the galaxy rotation problem based on a variation of Newton's Second Law of Dynamics at low accelerations. This would produce a large-scale variation of Newton's universal theory of gravity. A modification of Newton's theory would also imply a modification of general relativistic cosmology in as much as Newtonian cosmology is the limit of Friedman cosmology. While almost all astrophysicists today reject MOND in favor of dark matter, a small number of researchers continue to enhance it, recently incorporating Brans–Dicke theories into treatments that attempt to account for cosmological observations.

TeVeS

Tensor–vector–scalar gravity (TeVeS) is a proposed relativistic theory that is equivalent to Modified Newtonian dynamics (MOND) in the non-relativistic limit, which purports to explain the galaxy rotation problem without invoking dark matter. Originated by Jacob Bekenstein in 2004, it incorporates various dynamical and non-dynamical tensor fields, vector fields and scalar fields.

The break-through of TeVeS over MOND is that it can explain the phenomenon of gravitational lensing, a cosmic optical illusion in which matter bends light, which has been confirmed many times. A recent preliminary finding is that it can explain structure formation without CDM, but requiring a ~2eV massive neutrino (they are also required to fit some Clusters of galaxies, including the Bullet Cluster). However, other authors claim that TeVeS can't explain cosmic microwave background anisotropies and structure formation at the same time, i.e. ruling out those models at high significance.

f(R) Gravity

f(R) gravity is a family of theories that modify general relativity by defining a different function of the Ricci scalar. The simplest case is just the function being equal to the scalar; this is general relativity. As a consequence of introducing an arbitrary function, there may be freedom to explain the accelerated expansion and structure formation of the Universe without adding unknown forms of dark energy or dark matter. Some functional forms may be inspired by corrections arising from a quantum theory of gravity. f(R) gravity was first proposed in 1970 by Hans Adolph Buchdahl (although φ was used rather than f for the name of the arbitrary function). It has become an active field of research following work by Starobinsky on cosmic inflation. A wide range of phenomena can be produced from this theory by adopting different functions; however, many functional forms can now be ruled out on observational grounds, or because of pathological theoretical problems.

Steady State Theories

The Steady State theory extends the homogeneity assumption of the cosmological principle to reflect a homogeneity in time as well as in space. This "perfect cosmological principle" as it would come to be called asserted that the universe looks the same everywhere (on the large scale), the same as it always has and always will. This is in contrast to Lambda-CDM, in which the universe looked very different in the past and will look very different in the future. Steady State theory was proposed in 1948 by Fred Hoyle, Thomas Gold, Hermann Bondi and others. In order to maintain the perfect cosmological principle in an expanding universe, steady state cosmology had to posit a "matter-creation field" (the so-called C-field) that would insert matter into the universe in order to maintain a constant density.

The debate between the Big Bang and the Steady State models would happen for 15 years with camps roughly evenly divided until the discovery of the cosmic microwave background radiation. This radiation is a natural feature of the Big Bang model which demands a "time of last scattering" where photons decouple with baryonic matter. The Steady State model proposed that this radiation could be accounted for by so-called "integrated starlight" which was a background caused in part by Olbers' paradox in an infinite universe. In order to account for the uniformity of the background, steady state proponents posited a fog effect associated with microscopic iron particles that would scatter radio waves in such a manner as to produce an isotropic CMB. The proposed phenomena was whimsically named "cosmic iron whiskers" and served as the thermalization mechanism. The Steady State theory did not have the horizon problem of the Big Bang because it assumed an infinite amount of time was available for thermalizing the background.

As more cosmological data began to be collected, cosmologists began to realize that the Big Bang correctly predicted the abundance of light elements observed in the cosmos. What was a coincidental ratio of hydrogen to deuterium and helium in the steady state model was a feature of the Big Bang model. Additionally, detailed measurements of the CMB since the 1990s with the COBE, WMAP and Planck observations indicated that the spectrum of the background was closer to a blackbody than any other source in nature. The best integrated starlight models could predict was a thermalization to the level of 10% while the COBE satellite measured the deviation at one part in 10^5. After this dramatic discovery, the majority of cosmologists became convinced that the steady state theory could not explain the observed CMB properties.

Although the original steady state model is now considered to be contrary to observations (particularly the CMB) even by its one-time supporters, modifications of the steady state model has been proposed, including a model that envisions the universe as originating through many little bangs rather than one big bang (the so-called "quasi-steady state cosmology"). It supposes that the universe goes through periodic expansion and contraction phases, with a soft "rebound" in place of the Big Bang. Thus the Hubble Law is explained by the fact that the universe is currently in an expansion phase. Work continues on this model (most notably by Jayant V. Narlikar), although it has not gained widespread mainstream acceptance.

Anisotropic Universe

Isotropicity – the idea that the universe looks the same in all directions – is one of the core assumptions that enters into the FRW equations. In 2008 however, scientists working on Wilkinson Microwave Anisotropy Probe data claimed to have detected a 600–1000 km/s flow of clusters toward a 20-degree patch of sky between the constellations of Centaurus and Vela. They suggested that the motion may be a remnant of the influence of no-longer-visible regions of the universe prior to inflation. The detection is controversial, and other scientists have found that the universe is isotropic to a great degree.

Exotic Dark Matter and Dark Energy

In Lambda-CDM, dark matter is an extremely inert form of matter that does not interact with both ordinary matter (baryons) and light, but still exerts gravitational effects. To produce the large-scale structure we see today, dark matter is "cold" (the 'C' in Lambda-CDM), i.e. non-relativistic. Dark energy is an unknown form of energy that tends to accelerate the expansion of the universe. Both dark matter and dark energy have not been conclusively identified, and their exact nature is the subject of intense study. For example, scientists have hypothesized that dark matter could decay into dark energy, or that both dark matter and dark energy are different facets of the same underlying fluid. Other theories that aim to explain one or the other, such as warm dark matter and quintessence, also fall into this category.

Proposals based on Observational Skepticism

As the observational cosmology began to develop, certain astronomers began to offer alternative speculations regarding the interpretation of various phenomena that occasionally became parts of non-standard cosmologies.

Tired Light

Tired light theories challenge the common interpretation of Hubble's Law as a sign the universe is expanding. It was proposed by Fritz Zwicky in 1929. The basic proposal amounted to light losing energy ("getting tired") due to the distance it traveled rather than any metric expansion or physical recession of sources from observers. A traditional explanation of this effect was to attribute a dynamical friction to photons; the photons' gravitational interactions with stars and other material will progressively reduce their momentum, thus producing a redshift. Other proposals for explaining how photons could lose energy included the scattering of light by intervening material in a process similar to observed interstellar reddening. However, all these processes would also tend to blur images of distant objects, and no such blurring has been detected.

Traditional tired light has been found incompatible with the observed time dilation that is associated with the cosmological redshift. This idea is mostly remembered as a falsified alternative explanation for Hubble's law in most astronomy or cosmology discussions.

Dirac Large Numbers Hypothesis

The Dirac large numbers hypothesis uses the ratio of the size of the visible universe to the radius of quantum particle to predict the age of the universe. The coincidence of various ratios being close in order of magnitude may ultimately prove meaningless or the indication of a deeper connection between concepts in a future theory of everything. Nevertheless, attempts to use such ideas have been criticized as numerology.

Redshift Periodicity and Intrinsic Redshifts

Some astrophysicists were unconvinced that the cosmological redshifts are caused by universal cosmological expansion. Skepticism and alternative explanations began appearing in the scientific literature in the 1960s. In particular, Geoffrey Burbidge, William Tifft and Halton Arp were all observational astrophysicists who proposed that there were inconsistencies in the redshift observations of galaxies and quasars. The first two were famous for suggesting that there were periodicities in the redshift distributions of galaxies and quasars. Subsequent statistical analyses of redshift surveys, however, have not confirmed the existence of these periodicities.

During the quasar controversies of the 1970s, these same astronomers were also of the opinion that quasars exhibited high redshifts not due to their incredible distance but rather due to unexplained intrinsic redshift mechanisms that would cause the periodicities and cast doubt on the Big Bang. Arguments over how distant quasars were took the form of debates surrounding quasar energy production mechanisms, their light curves, and whether quasars exhibited any proper motion. Astronomers who believed quasars were not at cosmological distances argued that the Eddington luminosity set limits on how distant the quasars could be since the energy output required to explain the apparent brightness of cosmologically-distant quasars was far too high to be explainable by nuclear fusion alone. This objection was made moot by the improved models of gravity-powered accretion disks which for sufficiently dense material (such as black holes) can be more efficient at energy production than nuclear reactions. The controversy was laid to rest by the 1990s when evidence became available that observed quasars were actually the ultra-luminous cores of distant active galactic nuclei and that the major components of their redshift were in fact due to the Hubble flow.

Halton Arp in London.

Throughout his career, Halton Arp maintained that there were anomalies in his observations of quasars and galaxies, and that those anomalies served as a refutation of the Big Bang. In particular, Arp pointed out examples of quasars that were close to the line of sight of (relatively) nearby active, mainly Seyfert galaxies. These objects are now classified under the term active galactic nuclei (AGN), Arp criticized using such term on the ground that it isn't empirical. He claimed that clusters of quasars were in alignment around cores of these galaxies and that quasars, rather than being the cores of distant AGN, were actually much closer and were starlike-objects ejected from the centers of nearby galaxies with high intrinsic redshifts. Arp also contended that they gradually lost their non-cosmological redshift component and eventually evolved into full-fledged galaxies. This stands in stark contradiction to the accepted models of galaxy formation.

The biggest problem with Arp's analysis is that today there are hundreds of thousands of quasars with known redshifts discovered by various sky surveys. The vast majority of these quasars are not correlated in any way with nearby AGN. Indeed, with improved observing techniques, a number of host galaxies have been observed around quasars which indicates that those quasars at least really are at cosmological distances and are not the kind of objects Arp proposes. Arp's analysis, according to most scientists, suffers from being based on small number statistics and hunting for peculiar coincidences and odd associations. Unbiased samples of sources, taken from numerous galaxy surveys of the sky show none of the proposed 'irregularities', nor that any statistically significant correlations exist.

In addition, it is not clear what mechanism would be responsible for intrinsic redshifts or their gradual dissipation over time. It is also unclear how nearby quasars would explain some features in the spectrum of quasars which the standard model easily explains. In the standard cosmology, clouds of neutral hydrogen between the quasar and the earth create Lyman alpha absorption lines having different redshifts up to that of the quasar itself; this feature is called the Lyman-alpha forest. Moreover, in extreme quasars one can observe the absorption of neutral hydrogen which has not yet been reionized in a feature known as the Gunn–Peterson trough. Most cosmologists see this missing theoretical work as sufficient reason to explain the observations as either chance or error.

Halton Arp has proposed an explanation for his observations by a Machian "variable mass hypothesis". The variable-mass theory invokes constant matter creation from active galactic nuclei, which puts it into the class of steady-state theories. With the passing of Halton Arp, this cosmology has been relegated to a dismissed theory.

Nucleosynthesis Objections

One of the major successes of the Big Bang theory has been to provide a prediction that corresponds to the observations of the abundance of light elements in the universe. Along with the explanation provided for the Hubble's law and for the cosmic microwave background, this observation has proved very difficult for alternative theories to explain.

Theories which assert that the universe has an infinite age, including many of the theories described above, fail to account for the abundance of deuterium in the cosmos, because deuterium easily undergoes nuclear fusion in stars and there are no known astrophysical processes other than the Big Bang itself that can produce it in large quantities. Hence the fact that deuterium is not an extremely rare component of the universe suggests that the universe has a finite age.

Theories which assert that the universe has a finite life, but that the Big Bang did not happen, have problems with the abundance of helium-4. The observed amount of ^4He is far larger than the amount that should have been created via stars or any other known process. By contrast, the abundance of ^4He in Big Bang models is very insensitive to assumptions about baryon density, changing only a few percent as the baryon density changes by several orders of magnitude. The observed value of ^4He is within the range calculated.

Plasma Cosmology

In 1965, Hannes Alfvén proposed a "plasma cosmology" theory of the universe based in part on scaling observations of space plasma physics and experiments on plasmas in terrestrial laboratories to cosmological scales orders of magnitude greater. Taking matter–antimatter symmetry as a starting point, Alfvén together with Oskar Klein proposed the Alfvén-Klein cosmology model, based on the fact that since most of the local universe was composed of matter and not antimatter there may be large bubbles of matter and antimatter that would globally balance to equality. The difficulties with this model were apparent almost immediately. Matter–antimatter annihilation results in the production of high energy photons which were not observed. While it was possible that the local "matter-dominated" cell was simply larger than the observable universe, this proposition did not lend itself to observational tests.

Like the steady state theory, plasma cosmology includes a Strong Cosmological Principle which assumes that the universe is isotropic in time as well as in space. Matter is explicitly assumed to have always existed, or at least that it formed at a time so far in the past as to be forever beyond humanity's empirical methods of investigation.

While plasma cosmology has never had the support of most astronomers or physicists, a small number of plasma researchers have continued to promote and develop the approach, and publish in the special issues of the IEEE Transactions on Plasma Science. A few papers regarding plasma cosmology were published in other mainstream journals until the 1990s. Additionally, in 1991, Eric J. Lerner, an independent researcher in plasma physics and nuclear fusion, wrote a popular-level book supporting plasma cosmology called The Big Bang Never Happened. At that time there was renewed interest in the subject among the cosmological community along with other non-standard cosmologies. This was due to anomalous results reported in 1987 by Andrew Lange and Paul Richardson of UC Berkeley and Toshio Matsumoto of Nagoya University that indicated the cosmic microwave background might not have a blackbody spectrum.

However, the final announcement of COBE satellite data corrected the earlier contradiction of the Big Bang; the popularity of plasma cosmology has since fallen.

Plasma cosmology is a non-standard cosmology whose central postulate is that the dynamics of ionized gases and plasmas play important, if not dominant, roles in the physics of the universe beyond the Solar System. In contrast, the current observations and models of cosmologists and astrophysicists explain the formation, development, and evolution of astronomical bodies and large-scale structures in the universe as influenced by gravity (including its formulation in Einstein's theory of general relativity) and baryonic physics.

Some theoretical concepts about plasma cosmology originated with Hannes Alfvén, who tentatively proposed the use of plasma scaling to extrapolate the results of laboratory experiments and plasma physics observations and scale them over many orders of magnitude up to the largest observable objects in the universe.

Cosmologists and astrophysicists who have evaluated plasma cosmology reject it because it does not match the observations of astrophysical phenomena as well as current cosmological theory. Very few papers supporting plasma cosmology have appeared in the literature since the mid-1990s.

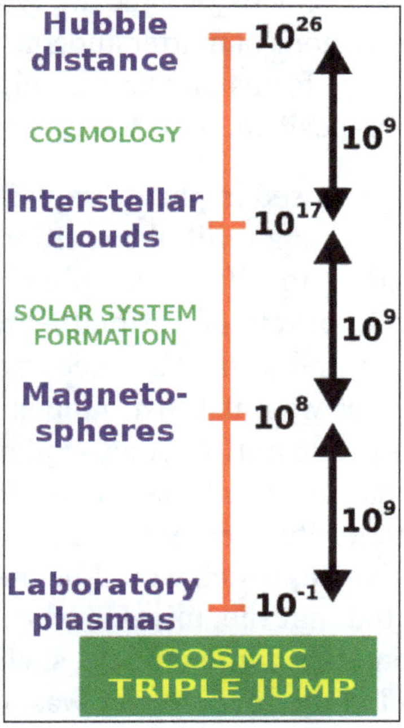

The term *plasma universe* is sometimes used as a synonym for plasma cosmology, as an alternative description of the plasma in the universe.

Hannes Alfvén suggested that scaling laboratory results can be extrapolated up to the scale of the universe. A scaling jump by a factor 10^9 was required to extrapolate to the magnetosphere, a second jump to extrapolate to galactic conditions, and a third jump to extrapolate to the Hubble distance.

Alfvén–klein Cosmology

In the 1960s, the theory behind plasma cosmology was introduced by Alfvén, a plasma expert who won the 1970 Nobel Prize in Physics for his work on magnetohydrodynamics (MHD). In 1971, Oskar Klein, a Swedish theoretical physicist, extended the earlier proposals and developed the Alfvén–Klein model of the universe, or “metagalaxy”, an earlier term used to refer to the empirically accessible part of the universe, rather than the entire universe including parts beyond our particle horizon. In this Alfvén–Klein cosmology, sometimes called Klein–Alfvén cosmology, the universe is made up of equal amounts of matter and antimatter with the boundaries between the regions of matter and antimatter being delineated by cosmic electromagnetic fields formed by double layers, thin regions comprising two parallel layers with opposite electrical charge. Interaction between these boundary regions would generate radiation, and this would form the plasma. Alfvén introduced the term ambiplasma for plasma made up of matter and antimatter and the double layers are thus formed of ambiplasma. According to

Alfvén, such an ambiplasma would be relatively long-lived as the component particles and antiparticles would be too hot and too low-density to annihilate each other rapidly. The double layers will act to repel clouds of opposite type, but combine clouds of the same type, creating ever-larger regions of matter and antimatter. The idea of ambiplasma was developed further into the forms of heavy ambiplasma (protons-antiprotons) and light ambiplasma (electrons-positrons).

Alfvén–Klein cosmology was proposed in part to explain the observed baryon asymmetry in the universe, starting from an initial condition of exact symmetry between matter and antimatter. According to Alfvén and Klein, ambiplasma would naturally form pockets of matter and pockets of antimatter that would expand outwards as annihilation between matter and antimatter occurred in the double layer at the boundaries. They concluded that we must just happen to live in one of the pockets that was mostly baryons rather than antibaryons, explaining the baryon asymmetry. The pockets, or bubbles, of matter or antimatter would expand because of annihilations at the boundaries, which Alfvén considered as a possible explanation for the observed expansion of the universe, which would be merely a local phase of a much larger history. Alfvén postulated that the universe has always existed due to causality arguments and the rejection of *ex nihilo* models, such as the Big Bang, as a stealth form of creationism. The exploding double layer was also suggested by Alfvén as a possible mechanism for the generation of cosmic rays, X-ray bursts and gamma-ray bursts.

In 1993, theoretical cosmologist Jim Peebles criticized Alfvén–Klein cosmology, writing that "there is no way that the results can be consistent with the isotropy of the cosmic microwave background radiation and X-ray backgrounds". In his book he also showed that Alfvén's models do not predict Hubble's law, the abundance of light elements, or the existence of the cosmic microwave background. A further difficulty with the ambiplasma model is that matter–antimatter annihilation results in the production of high energy photons, which are not observed in the amounts predicted. While it is possible that the local "matter-dominated" cell is simply larger than the observable universe, this proposition does not lend itself to observational tests.

Plasma Cosmology and the Study of Galaxies

Hannes Alfvén from the 1960s to 1980s argued that plasma played an important if not dominant role in the universe because electromagnetic forces are far more important than gravity when acting on interplanetary and interstellar charged particles. He further hypothesized that they might promote the contraction of interstellar clouds and may even constitute the main mechanism for contraction, initiating star formation. The current standard view is that magnetic fields can hinder collapse, that large-scale Birkeland currents have not been observed, and that the length scale for charge neutrality is predicted to be far smaller than the relevant cosmological scales.

In the 1980s and 1990s, Alfvén and Anthony Peratt, a plasma physicist at Los Alamos National Laboratory, outlined a program they called the "plasma universe". In plasma universe proposals, various plasma physics phenomena were associated with astrophysical observations and were used to explain extant mysteries and problems outstanding in astrophysics in the 1980s and 1990s. In various venues, Peratt profiled what he characterized as an alternative viewpoint to the mainstream models applied in astrophysics and cosmology.

For example, Peratt proposed that the mainstream approach to galactic dynamics which relied on gravitational modeling of stars and gas in galaxies with the addition of dark matter was overlooking a possibly major contribution from plasma physics. He mentions laboratory experiments of Winston H. Bostick in the 1950s that created plasma discharges that looked like galaxies. Perrat conducted computer simulations of colliding plasma clouds that he reported also mimicked the shape of galaxies. Peratt proposed that galaxies formed due to plasma filaments joining in a z-pinch, the filaments starting 300,000 light years apart and carrying Birkeland currents of 10^{18} amperes. Peratt also reported simulations he did showing emerging jets of material from the central buffer region that he compared to quasars and active galactic nuclei occurring without supermassive black holes. Peratt proposed a sequence for galaxy evolution: "the transition of double radio galaxies to radioquasars to radioquiet QSO's to peculiar and Seyfert galaxies, finally ending in spiral galaxies". He also reported that flat galaxy rotation curves were simulated without dark matter. At the same time Eric Lerner, an independent plasma researcher and supporter of Peratt's ideas, proposed a plasma model for quasars based on a dense plasma focus.

Comparison to Mainstream Astrophysics

Standard astronomical modeling and theories attempt to incorporate all known physics into descriptions and explanations of observed phenomena, with gravity playing a dominant role on the largest scales as well as in celestial mechanics and dynamics. To that end, both Keplerian orbits and Einstein's general theory of relativity are generally used as the underlying frameworks for modeling astrophysical systems and structure formation, while high-energy astronomy and particle physics in cosmology additionally appeal to electromagnetic processes including plasma physics and radiative transfer to explain relatively small scale energetic processes observed in the x-rays and gamma rays. Due to overall charge neutrality, plasma physics does not provide for very long-range interactions in astrophysics even while much of the matter in the universe is plasma.

Proponents of plasma cosmology claim electrodynamics is as important as gravity in explaining the structure of the universe, and speculate that it provides an alternative explanation for the evolution of galaxies and the initial collapse of interstellar clouds. In particular plasma cosmology is claimed to provide an alternative explanation for the flat rotation curves of spiral galaxies and to do away with the need for dark matter in galaxies and with the need for supermassive black holes in galaxy centres to power quasars and active galactic nuclei. However, theoretical analysis shows that "many scenarios for the generation of seed magnetic fields, which rely on the survival and sustainability of currents at early times of the universe are disfavored", i.e. Birkeland currents of the magnitude needed (10^{18} amps over scales of megaparsecs) for galaxy formation do not exist. Additionally, many of the issues that were mysterious in the 1980s and 1990s, including discrepancies relating to the cosmic microwave background and the nature of quasars, have been solved with more evidence that, in detail, provides a distance and time scale for the universe.

Some of the places where plasma cosmology supporters are most at odds with standard explanations include the need for their models to have light element production without Big Bang nucleosynthesis, which, in the context of Alfvén–Klein cosmology, has been shown to produce excessive X-rays and gamma rays beyond that observed. Plasma cosmology proponents have made further proposals to explain light element abundances, but the attendant issues have not been fully addressed. In 1995 Eric Lerner published his alternative explanation for the cosmic microwave background radiation (CMB). He argued that his model explained the fidelity of the CMB spectrum to that of a black body and the low level of anisotropies found, even while the level of isotropy at $1:10^5$ is not accounted for to that precision by any alternative models. Additionally, the sensitivity and resolution of the measurement of the CMB anisotropies was greatly advanced by WMAP and the Planck satellite and the statistics of the signal were so in line with the predictions of the Big Bang model, that the CMB has been heralded as a major confirmation of the Big Bang model to the detriment of alternatives. The acoustic peaks in the early universe are fit with high accuracy by the predictions of the Big Bang model, and, to date, there has never been an attempt to explain the detailed spectrum of the anisotropies within the framework of plasma cosmology or any other alternative cosmological model.

Cosmological Principle

In modern physical cosmology, the cosmological principle is the notion that the spatial distribution of matter in the universe is homogeneous and isotropic when viewed on a large enough scale, since the forces are expected to act uniformly throughout the universe, and should, therefore, produce no observable irregularities in the large-scale structuring over the course of evolution of the matter field that was initially laid down by the Big Bang.

Astronomer William Keel explains:

"The cosmological principle is usually stated formally as 'Viewed on a sufficiently large scale, the properties of the universe are the same for all observers.' This amounts to the strongly philosophical statement that the part of the universe which we can see is a fair sample, and that the same physical laws apply throughout. In essence, this in a sense says that the universe is knowable and is playing fair with scientists."

The cosmological principle depends on a definition of "observer," and contains an implicit qualification and two testable consequences.

"Observers" means any observer at any location in the universe, not simply any human observer at any location on Earth: as Andrew Liddle puts it, "the cosmological principle means that the universe looks the same whoever and wherever you are."

The qualification is that variation in physical structures can be overlooked, provided this does not imperil the uniformity of conclusions drawn from observation: The Sun is different from the Earth, our galaxy is different from a black hole, some galaxies advance toward rather than recede from us, and the universe has a "foamy" texture of galaxy clusters and voids, but none of these different structures appears to violate the basic laws of physics.

The two testable structural consequences of the cosmological principle are homogeneity and isotropy. Homogeneity means that the same observational evidence is available to observers at different locations in the universe ("the part of the universe which we can see is a fair sample"). Isotropy means that the same observational evidence is available by looking in any direction in the universe ("the same physical laws apply throughout". The principles are distinct but closely related, because a universe that appears isotropic from any two (for a spherical geometry, three) locations must also be homogeneous.

The cosmological principle is first clearly asserted in the *Philosophiæ Naturalis Principia Mathematica* of Isaac Newton. In contrast to earlier classical or medieval cosmologies, in which Earth rested at the center of universe, Newton conceptualized the Earth as a sphere in orbital motion around the Sun within an empty space that extended uniformly in all directions to immeasurably large distances. He then showed, through a series of mathematical proofs on detailed observational data of the motions of planets and comets, that their motions could be explained by a single principle of "universal gravitation" that applied as well to the orbits of the Galilean moons around Jupiter, the Moon around the Earth, the Earth around the Sun, and to falling bodies on Earth. That is, he asserted the equivalent material nature of all bodies within the Solar System, the identical nature of the Sun and distant stars and thus the uniform extension of the physical laws of motion to a great distance beyond the observational location of Earth itself.

Implications of Cosmological Principle

Observations show that more distant galaxies are closer together and have lower content of chemical elements heavier than lithium. Applying the cosmological principle, this suggests that heavier elements were not created in the Big Bang but were produced by nucleosynthesis in giant stars and expelled across a series of supernovae explosions and new star formation from the supernovae remnants, which means heavier elements would accumulate over time. Another observation is that the furthest galaxies (earlier time) are often more fragmentary, interacting and unusually shaped than local galaxies (recent time), suggesting evolution in galaxy structure as well.

A related implication of the cosmological principle is that the largest discrete structures in the universe are in mechanical equilibrium. Homogeneity and isotropy of matter at the largest scales would suggest that the largest discrete structures are parts of a single indiscrete form, like the crumbs which make up the interior of a cake. At extreme cosmological distances, the property of mechanical equilibrium in surfaces lateral to the line of sight can be empirically tested; however, under the assumption of the cosmological principle, it cannot be detected parallel to the line of sight.

Cosmologists agree that in accordance with observations of distant galaxies, a universe must be non-static if it follows the cosmological principle. In 1923, Alexander Friedmann set out a variant of Einstein's equations of general relativity that describe the dynamics of a homogeneous isotropic universe. Independently, Georges Lemaître derived in 1927 the equations of an expanding universe from the General Relativity equations. Thus, a non-static universe is also implied, independent of observations of distant galaxies, as the result of applying the cosmological principle to general relativity.

Criticism of Cosmological Principle

Karl Popper criticized the cosmological principle on the grounds that it makes "our *lack* of knowledge a principle of *knowing something*". He summarized his position as:

"The 'cosmological principles' were, I fear, dogmas that should not have been proposed."

Observations of Cosmological Principle

Although the universe is inhomogeneous at smaller scales, it *is* statistically homogeneous on scales larger than 250 million light years. The cosmic microwave background is isotropic, that is to say that its intensity is about the same whichever direction we look at.

However, recent findings have called this view into question. Data from the Planck Mission shows hemispheric bias in 2 respects: One with respect to average temperature (i.e. temperature fluctuations), the second with respect to larger variations in the degree of perturbations (i.e. densities). Therefore, the European Space Agency has concluded that these anisotropies are, in fact, statistically significant and can no longer be ignored.

Inconsistencies of Cosmological Principle

The *cosmological principle* implies that at a sufficiently large scale, the universe is homogeneous. This means that different places will appear similar to one another, so sufficiently large structures cannot exist. Yadav and his colleagues have suggested a maximum scale of 260/h Mpc for structures within the universe according to this heuristic. Other authors have suggested values as low as 60/h Mpc. Yadav's calculation suggests that the maximum size of a structure can be about 370 Mpc.

A number of observations conflict with predictions of maximal structure sizes:

- The Clowes–Campusano LQG, discovered in 1991, has a length of 580 Mpc, and is marginally larger than the consistent scale.
- The Sloan Great Wall, discovered in 2003, has a length of 423 Mpc, which is only just consistent with the cosmological principle.
- U1.11, a large quasar group discovered in 2011, has a length of 780 Mpc, and is two times larger than the upper limit of the homogeneity scale.
- The Huge-LQG, discovered in 2012, is three times longer than, and twice as wide as is predicted possible according to these current models, and so challenges our understanding of the universe on large scales.
- In November 2013, a new structure 10 billion light years away measuring 2000-3000 Mpc (more than seven times that of the SGW) has been discovered, the Hercules–Corona Borealis Great Wall, putting further doubt on the validity of the cosmological principle.

In September 2016, however, studies of the expansion of the Universe that have used data taken by the *Planck* mission show it to be highly isotropical, reinforcing the cosmological principle.

Perfect Cosmological Principle

The perfect cosmological principle is an extension of the cosmological principle, and states that the universe is homogeneous and isotropic in space *and* time. In this view the universe looks the same everywhere (on the large scale), the same as it always has and always will. The perfect cosmological principle underpins Steady State theory and emerges from chaotic inflation theory.

Comoving and Proper Distances

In standard cosmology, comoving distance and proper distance are two closely related distance measures used by cosmologists to define distances between objects. *Proper distance* roughly corresponds to where a distant object would be at a specific moment of cosmological time, which can change over time due to the expansion of the universe. *Comoving distance* factors out the expansion of the universe, giving a distance that does not change in time due to the expansion of space (though this may change due to other, local factors, such as the motion of a galaxy within a cluster). Comoving distance and proper distance are defined to be equal at the present time; therefore, the ratio of proper distance to comoving distance now is 1. At other times, the scale factor differs from 1. The Universe's expansion results in the proper distance changing, while the comoving distance is unchanged by this expansion because it is the proper distance divided by that scale factor.

Comoving Coordinates

The evolution of the universe and its horizons in comoving distances. The x-axis is distance, in billions of light years; the left-hand y-axis is time, in billions of years since the Big Bang; the right-hand y-axis is the scale factor. This model of the universe includes dark energy which causes an accelerating expansion after a certain point in time, and results in an event horizon beyond which we can never see.

Although general relativity allows one to formulate the laws of physics using arbitrary coordinates, some coordinate choices are more natural or easier to work with. Comoving coordinates are an example of such a natural coordinate choice. They assign constant spatial coordinate values to observers who perceive the universe as isotropic. Such observers are called "comoving" observers because they move along with the Hubble flow.

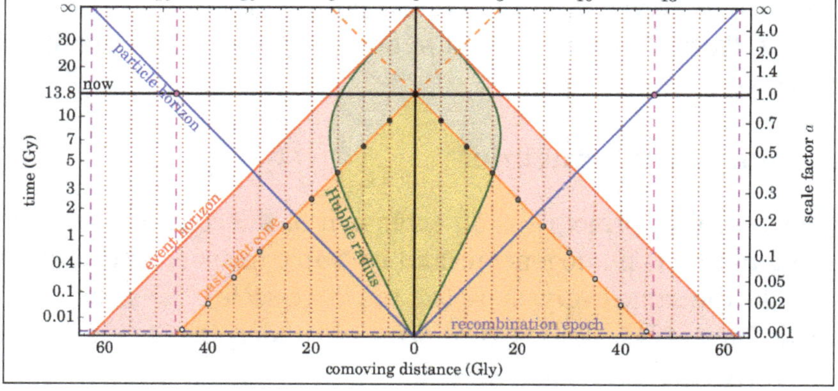

A comoving observer is the only observer that will perceive the universe, including the cosmic microwave background radiation, to be isotropic. Non-comoving observers will see regions of the sky systematically blue-shifted or red-shifted. Thus isotropy, particularly isotropy of the cosmic microwave background radiation, defines a special local frame of reference called the comoving frame. The velocity of an observer relative to the local comoving frame is called the peculiar velocity of the observer.

Most large lumps of matter, such as galaxies, are nearly comoving, so that their peculiar velocities (owing to gravitational attraction) are low.

The comoving time coordinate is the elapsed time since the Big Bang according to a clock of a comoving observer and is a measure of cosmological time. The comoving spatial coordinates tell where an event occurs while cosmological time tells when an event occurs. Together, they form a complete coordinate system, giving both the location and time of an event.

Space in comoving coordinates is usually referred to as being "static", as most bodies on the scale of galaxies or larger are approximately comoving, and comoving bodies have static, unchanging comoving coordinates. So for a given pair of comoving galaxies, while the proper distance between them would have been smaller in the past and will become larger in the future due to the expansion of space, the comoving distance between them remains *constant* at all times.

The expanding Universe has an increasing scale factor which explains how constant comoving distances are reconciled with proper distances that increase with time.

Comoving Distance and Proper Distance

Comoving distance is the distance between two points measured along a path defined at the present cosmological time. For objects moving with the Hubble flow, it is deemed to remain constant in time. The comoving distance from an observer to a distant object (e.g. galaxy) can be computed by the following formula (derived using the Friedmann-Lemaître-Robertson-Walker metric):

$$\chi = \int_{t_e}^{t} c \frac{dt'}{a(t')}$$

where $a(t')$ is the scale factor, t_e is the time of emission of the photons detected by the observer, t is the present time, and c is the speed of light in vacuum.

Despite being an integral over time, this expression gives the correct distance that would be measured by a hypothetical tape measure at fixed time t, i.e. the "proper distance" after accounting for the time-dependent *comoving speed of light* via the inverse scale factor term $1/a(t')$ in the integrand. By "comoving speed of light", we mean the velocity of light *through* comoving coordinates $[c/a(t')]$ which is time-dependent even though *locally*, at any point along the null geodesic of the light particles, an observer in an inertial frame always measures the speed of light as c in accordance with special relativity.

Many use the symbol χ for the comoving distance. However, this χ must be distinguished from the coordinate distance r in the commonly used comoving coordinate system for a FLRW universe where the metric takes the form (in reduced-circumference polar coordinates, which only works half-way around a spherical universe):

$$ds^2 = -c^2\,d\tau^2 = -c^2\,dt^2 + a(t)^2\left(\frac{dr^2}{1-\kappa r^2} + r^2\left(d\theta^2 + \sin^2\theta\,d\phi^2\right)\right).$$

In this case the comoving coordinate distance r is related to χ by:

$$\chi = \begin{cases} |\kappa|^{-1/2}\sinh^{-1}\sqrt{|\kappa|}r, & \text{if } \kappa < 0 \text{ (a negatively curved 'hyperbolic' universe)} \\ r, & \text{if } \kappa = 0 \text{ (a spatially flat universe)} \\ |\kappa|^{-1/2}\sin^{-1}\sqrt{|\kappa|}r, & \text{if } \kappa > 0 \text{ (a positively curved 'spherical' universe)} \end{cases}$$

Most textbooks and research papers define the comoving distance between comoving observers to be a fixed unchanging quantity independent of time, while calling the dynamic, changing distance between them "proper distance". On this usage, comoving and proper distances are numerically equal at the current age of the universe, but will differ in the past and in the future; if the comoving distance to a galaxy is denoted χ, the proper distance $d(t)$ at an arbitrary time t is simply given by $d(t) = a(t)\chi$ where $a(t)$ is the scale factor. The proper distance $d(t)$ between two galaxies at time t is just the distance that would be measured by rulers between them at that time.

Uses of the Proper Distance

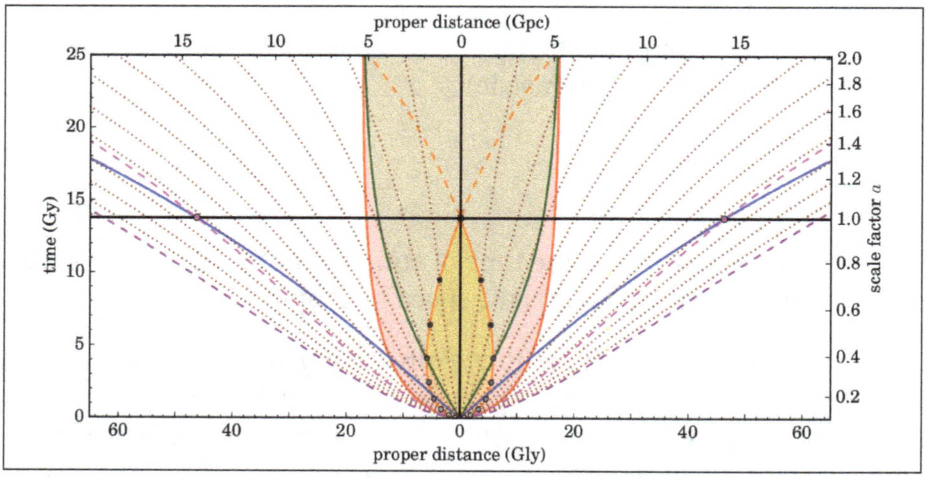

The evolution of the universe and its horizons in proper distances. The x-axis is distance, in billions of light years; the left-hand y-axis is time, in billions of years since the Big Bang; the right-hand y-axis is the scale factor. This is the same model as in the earlier figure, with dark energy and an event horizon.

Cosmological time is identical to locally measured time for an observer at a fixed comoving spatial position, that is, in the local comoving frame. Proper distance is also equal to the locally measured distance in the comoving frame for nearby objects. To measure the proper distance between two distant objects, one imagines that one has many comoving observers in a straight line between the two objects, so that all of the observers are close to each other, and form a chain between the two distant objects. All of these observers must have the same cosmological time. Each observer measures their distance to the nearest observer in the chain, and the length of the chain, the sum of distances between nearby observers, is the total proper distance.

It is important to the definition of both comoving distance and proper distance in the cosmological sense (as opposed to proper length in special relativity) that all observers have the same cosmological age. For instance, if one measured the distance along a straight line or spacelike geodesic between the two points, observers situated between the two points would have different cosmological ages when the geodesic path crossed their own world lines, so in calculating the distance along this geodesic one would not be correctly measuring comoving distance or cosmological proper distance. Comoving and proper distances are not the same concept of distance as the concept of distance in special relativity. This can be seen by considering the hypothetical case of a universe empty of mass, where both sorts of distance can be measured. When the density of mass in the FLRW metric is set to zero (an empty 'Milne universe'), then the cosmological coordinate system used to write this metric becomes a non-inertial coordinate system in the Minkowski spacetime of special relativity where surfaces of constant Minkowski proper-time τ appear as hyperbolas in the Minkowski diagram from the perspective of an inertial frame of reference. In this case, for two events which are simultaneous according to the cosmological time coordinate, the value of the cosmological proper distance is not equal to the value of the proper length between these same events, which would just be the distance along a straight line between the events in a Minkowski diagram (and a straight line is a geodesic in flat Minkowski spacetime), or the coordinate distance between the events in the inertial frame where they are simultaneous.

If one divides a change in proper distance by the interval of cosmological time where the change was measured (or takes the derivative of proper distance with respect to cosmological time) and calls this a "velocity", then the resulting "velocities" of galaxies or quasars can be above the speed of light, c. This apparent superluminal expansion is not in conflict with special or general relativity, and is a consequence of the particular definitions used in physical cosmology. Even light itself does not have a "velocity" of *c* in this sense; the total velocity of any object can be expressed as the sum $v_{\text{tot}} = v_{\text{rec}} + v_{\text{pec}}$ where v_{rec} is the recession velocity due to the expansion of the universe (the velocity given by Hubble's law) and v_{pec} is the "peculiar velocity" measured by local observers (with $v_{\text{rec}} = \dot{a}(t)\chi(t)$ and $v_{\text{pec}} = a(t)\dot{\chi}(t)$, the dots indicating a first derivative), so for light v_{pec} is equal to c ($-c$ if the light is emitted towards our position at the origin and +c if emitted away from us) but the total velocity v_{tot} is generally different from c. Even in special relativity the coordinate speed of light is only guaranteed to be c in an inertial frame; in a non-inertial frame the coordinate speed may be different from *c*. In general relativity no coordinate system on a large region of curved spacetime is "inertial", but in the local neighborhood of any point in curved spacetime we can define a "local inertial frame" in which the local speed of light is *c* and in which massive objects such as stars and galaxies always have a local speed smaller than *c*. The cosmological definitions used to define the velocities of distant objects are coordinate-dependent – there is no general coordinate-independent definition of velocity between distant objects in general relativity. The issue of how best to describe and popularize the apparent superluminal expansion of the universe has caused a minor amount of controversy.

Short Distances vs. Long Distances

Within small distances and short trips, the expansion of the universe during the trip can be ignored. This is because the travel time between any two points for a non-relativistic moving particle will just be the proper distance (that is, the comoving distance measured using the scale factor of the universe at the time of the trip rather than the scale factor "now") between those points divided by the velocity of the particle. If the particle is moving at a relativistic velocity, the usual relativistic corrections for time dilation must be made.

Cosmic Neutrino Background

The cosmic neutrino background (CNB or CνB) is the universe's background particle radiation composed of neutrinos. They are sometimes known as relic neutrinos.

The CNB is a relic of the Big Bang; while the cosmic microwave background radiation (CMB) dates from when the universe was 379,000 years old, the CNB decoupled (separated) from matter when the universe was just one second old. It is estimated that today, the CNB has a temperature of roughly 1.95 K.

As neutrinos rarely interact with matter, these neutrinos still exist today. They have a very low energy, around 10^{-4} to 10^{-6} eV. Even high energy neutrinos are notoriously difficult to detect, and the CνB has energies around 10^{10} times smaller, so the CνB may not be directly observed in detail for many years, if at all. However, Big Bang cosmology makes many predictions about the CνB, and there is very strong indirect evidence that the CνB exists.

Derivation of the CνB Temperature

Given the temperature of the CMB, the temperature of the CνB can be estimated. Before neutrinos decoupled from the rest of matter, the universe primarily consisted of neutrinos, electrons, positrons, and photons, all in thermal equilibrium with each other. Once the temperature dropped to approximately 2.5 MeV, the neutrinos decoupled from the rest of matter. Despite this decoupling, neutrinos and photons remained at the same temperature as the universe expanded. However, when the temperature dropped below the mass of the electron, most electrons and positrons annihilated, transferring their heat and entropy to photons, and thus increasing the temperature of the photons. So the ratio of the temperature of the photons before and after the electron-positron annihilation is the same as the ratio of the temperature of the neutrinos and the photons today. To find this ratio, we assume that the entropy of the universe was approximately conserved by the electron-positron annihilation. Then using:

$$\sigma \propto gT^3,$$

where σ is the entropy, g is the effective degrees of freedom and T is the temperature, we find that:

$$\left(\frac{g_0}{g_1}\right)^{\frac{1}{3}} = \frac{T_1}{T_0},$$

where T_0 denotes the temperature before the electron-positron annihilation and T_1 denotes after. The factor g_0 is determined by the particle species:

- 2 for photons, since they are massless bosons.
- 2 × (7/8) each for electrons and positrons, since they are fermions.

g_1 is just 2 for photons. So:

$$\frac{T_\nu}{T_\gamma} = \left(\frac{2}{2+2\times 7/8+2\times 7/8}\right)^{\frac{1}{3}} = \left(\frac{4}{11}\right)^{\frac{1}{3}}.$$

Given the current value of $T_\gamma = 2.725$ K, it follows that $T_\nu \approx 1.95$ K.

The above discussion is valid for massless neutrinos, which are always relativistic. For neutrinos with a non-zero rest mass, the description in terms of a temperature is no longer appropriate after they become non-relativistic; i.e., when their thermal energy $3/2\ kT_\nu$ falls below the rest mass energy $m_\nu c^2$. Instead, in this case one should rather track their energy density, which remains well-defined.

Indirect Evidence for the CνB

Relativistic neutrinos contribute to the radiation energy density of the universe ρ_R, typically parameterized in terms of the effective number of neutrino species N_ν:

$$\rho_R = \frac{\pi^2}{15} T_\gamma^4 (1+z)^4 \left[1 + \frac{7}{8} N_\nu \left(\frac{4}{11}\right)^{\frac{4}{3}}\right],$$

where z denotes the redshift. The first term in the square brackets is due to the CMB, the second comes from the CνB. The Standard Model with its three neutrino species predicts a value of $N_\nu \simeq 3.046$, including a small correction caused by a non-thermal distortion of the spectra during e^+-e^--annihilation. The radiation density had a major impact on various physical processes in the early universe, leaving potentially detectable imprints on measurable quantities, thus allowing us to infer the value of N_ν from observations.

Big Bang Nucleosynthesis

Due to its effect on the expansion rate of the universe during Big Bang nucleosynthesis (BBN), the theoretical expectations for the primordial abundances of light elements depend on N_ν. Astrophysical measurements of the primordial ^{4}He and ^{2}D abundances lead to a value of $N_\nu = 3.14^{+0.70}_{-0.65}$ at 68% c.l., in very good agreement with the Standard Model expectation.

CMB Anisotropies and Structure Formation

The presence of the CνB affects the evolution of CMB anisotropies as well as the growth of matter perturbations in two ways: due to its contribution to the radiation density of the universe (which determines for instance the time of matter-radiation equality), and due to the neutrinos' anisotropic stress which dampens the acoustic oscillations of the spectra. Additionally, free-streaming massive neutrinos suppress the growth of structure on small scales. The WMAP spacecraft's five-year data combined with type Ia supernova data and information about the baryon acoustic oscillation scale yielded $N_\nu = 4.34^{+0.88}_{-0.86}$ at 68% c.l., providing an independent confirmation of the BBN constraints. The Planck spacecraft collaboration has published the tightest bound to date on the effective number of neutrino species, at $N_\nu = 3.15 \pm 0.23$.

Indirect Evidence from Phase Changes to the Cosmic Microwave Background (CMB)

Big Bang cosmology makes many predictions about the CνB, and there is very strong indirect evidence that the cosmic neutrino background exists, both from Big Bang nucleosynthesis predictions of the helium abundance, and from anisotropies in the cosmic microwave background. One of these predictions is that neutrinos will have left a subtle imprint on the cosmic microwave background (CMB). It is well known that the CMB has irregularities. Some of the CMB fluctuations were roughly regularly spaced, because of the effect of baryon acoustic oscillation. In theory, the decoupled neutrinos should have had a very slight effect on the phase of the various CMB fluctuations.

In 2015, it was reported that such shifts had been detected in the CMB. Moreover, the fluctuations corresponded to neutrinos of almost exactly the temperature predicted by

Big Bang theory (1.96 ± 0.02 K compared to a prediction of 1.95 K), and exactly three types of neutrino, the same number of neutrino flavours currently predicted by the Standard Model.

Prospects for the Direct Detection of the CνB

Confirmation of the existence of these relic neutrinos may only be possible by directly detecting them using experiments on Earth. This will be difficult as the neutrinos which make up the CνB are non-relativistic, in addition to interacting only weakly with normal matter, and so any effect they have in a detector will be hard to identify. One proposed method of direct detection of the CνB is to use capture of cosmic relic neutrinos on tritium i.e. ^{3}H, leading to an induced form of beta decay. The neutrinos of the CνB would lead to the production of electrons via the reaction $\nu + {}^3\mathrm{H} \rightarrow {}^3\mathrm{He} + e^-$, while the main background comes from electrons produced via natural beta decay ${}^3\mathrm{H} \rightarrow {}^3\mathrm{He} + e^- + \bar{\nu}$. These electrons would be detected by the experimental apparatus in order to measure the size of the CνB. The latter source of electrons is far more numerous, however their maximum energy is smaller than the average energy of the CνB-electrons by twice the average neutrino mass. Since this mass is tiny, of the order of a few eVs or less, such a detector must have an excellent energy resolution in order to separate the signal from the background. One such proposed experiment is called PTOLEMY, which will be made up of 100 g of tritium target. The detector should be ready by 2022.

Friedmann Equations

The Friedmann equations are a set of equations in physical cosmology that govern the expansion of space in homogeneous and isotropic models of the universe within the context of general relativity. They were first derived by Alexander Friedmann in 1922 from Einstein's field equations of gravitation for the Friedmann–Lemaître–Robertson–Walker metric and a perfect fluid with a given mass density ρ and pressure ρ. The equations for negative spatial curvature were given by Friedmann in 1924.

Assumptions

The Friedmann equations start with the simplifying assumption that the universe is spatially homogeneous and isotropic, i.e. the cosmological principle; empirically, this is justified on scales larger than ~100 Mpc. The cosmological principle implies that the metric of the universe must be of the form:

$$ds^2 = a(t)^2\, ds_3^2 - c^2\, dt^2$$

where ds_3^2 is a three-dimensional metric that must be one of (a) flat space, (b) a sphere of constant positive curvature or (c) a hyperbolic space with constant negative curvature. The parameter k discussed below takes the value 0, 1, −1, or the Gaussian curvature, in these three cases respectively. It is this fact that allows us to sensibly speak of a "scale factor", $a(t)$.

Einstein's equations now relate the evolution of this scale factor to the pressure and energy of the matter in the universe. From FLRW metric we compute Christoffel symbols, then the Ricci tensor. With the stress–energy tensor for a perfect fluid, we substitute them into Einstein's field equations:

There are two independent Friedmann equations for modeling a homogeneous, isotropic universe. The first is:

$$\frac{\dot{a}^2 + kc^2}{a^2} = \frac{8\pi G\rho + \Lambda c^2}{3}$$

which is derived from the 00 component of Einstein's field equations. The second is:

$$\frac{\ddot{a}}{a} = -\frac{4\pi G}{3}\left(\rho + \frac{3p}{c^2}\right) + \frac{\Lambda c^2}{3}$$

which is derived from the first together with the trace of Einstein's field equations. a is the scale factor, $H \equiv \frac{\dot{a}}{a}$ is the Hubble parameter. G, Λ, and c are universal constants (G is Newton's gravitational constant, Λ is the cosmological constant, and c is the speed of light in vacuum). k is constant throughout a particular solution, but may vary from one solution to another. a, H, ρ, and p are functions of time. ρ, and p are the density and pressure, respectively. $\frac{k}{a^2}$ is the spatial curvature in any time-slice of the universe; it is equal to one-sixth of the spatial Ricci curvature scalar R since $R = \frac{6}{c^2 a^2}(\ddot{a}a + \dot{a}^2 + kc^2)$ in the Friedmann model. We see that in the Friedmann equations, a(t) depends only on ρ, p, Λ, and intrinsic curvature k. It does not depend on which coordinate system we chose for spatial slices. There are two commonly used choices for a and k which describe the same physics:

- k = +1, 0 or −1 depending on whether the shape of the universe is a closed 3-sphere, flat (i.e. Euclidean space) or an open 3-hyperboloid, respectively. If k = +1, then a is the radius of curvature of the universe. If k = 0, then a may be fixed to any arbitrary positive number at one particular time. If k = −1, then (loosely speaking) one can say that $i \cdot a$ is the radius of curvature of the universe.
- a is the scale factor which is taken to be 1 at the present time. k is the spatial curvature when $a = 1$ (i.e. today). If the shape of the universe is hyperspherical

and R_t is the radius of curvature (R_0 in the present-day), then $a = R_t / R_0$. If k is positive, then the universe is hyperspherical. If k is zero, then the universe is flat. If k is negative, then the universe is hyperbolic.

Using the first equation, the second equation can be re-expressed as:

$$\dot{\rho} = -3H\left(\rho + \frac{p}{c^2}\right),$$

which eliminates Λ and expresses the conservation of mass-energy $T^{\alpha\beta}{}_{;\beta} = 0$.

These equations are sometimes simplified by replacing:

$$\rho \to \rho - \frac{\Lambda c^2}{8\pi G}$$

$$p \to p + \frac{\Lambda c^4}{8\pi G}$$

to give:

$$H^2 = \left(\frac{\dot{a}}{a}\right)^2 = \frac{8\pi G}{3}\rho - \frac{kc^2}{a^2}$$

$$\dot{H} + H^2 = \frac{\ddot{a}}{a} = -\frac{4\pi G}{3}\left(\rho + \frac{3p}{c^2}\right).$$

The simplified form of the second equation is invariant under this transformation.

The Hubble parameter can change over time if other parts of the equation are time dependent (in particular the mass density, the vacuum energy, or the spatial curvature). Evaluating the Hubble parameter at the present time yields Hubble's constant which is the proportionality constant of Hubble's law. Applied to a fluid with a given equation of state, the Friedmann equations yield the time evolution and geometry of the universe as a function of the fluid density.

Some cosmologists call the second of these two equations the Friedmann acceleration equation and reserve the term *Friedmann equation* for only the first equation.

Density Parameter

The density parameter, Ω, is defined as the ratio of the actual (or observed) density ρ to the critical density ρ_c of the Friedmann universe. The relation between the actual density and the critical density determines the overall geometry of the universe; when they are equal, the geometry of the universe is flat (Euclidean). In earlier models, which did not include a cosmological constant term, critical density was initially defined as the watershed point between an expanding and a contracting Universe.

To date, the critical density is estimated to be approximately five atoms (of monatomic hydrogen) per cubic metre, whereas the average density of ordinary matter in the Universe is believed to be 0.2–0.25 atoms per cubic metre.

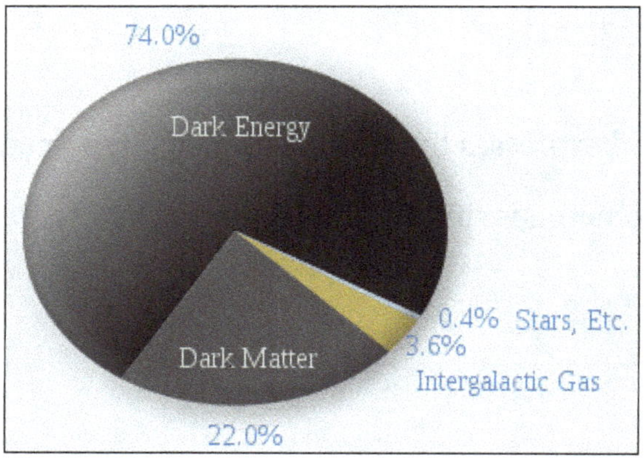

Estimated relative distribution for components of the energy density of the universe. Dark energy dominates the total energy (74%) while dark matter (22%) constitutes most of the mass. Of the remaining baryonic matter (4%), only one tenth is compact. In February 2015, the European-led research team behind the Planck cosmology probe released new data refining these values to 4.9% ordinary matter, 25.9% dark matter and 69.1% dark energy.

A much greater density comes from the unidentified dark matter; both ordinary and dark matter contribute in favour of contraction of the universe. However, the largest part comes from so-called dark energy, which accounts for the cosmological constant term. Although the total density is equal to the critical density (exactly, up to measurement error), the dark energy does not lead to contraction of the universe but rather may accelerate its expansion. Therefore, the universe will likely expand forever.

An expression for the critical density is found by assuming Λ to be zero (as it is for all basic Friedmann universes) and setting the normalised spatial curvature, k, equal to zero. When the substitutions are applied to the first of the Friedmann equations we find:

$$\rho_c = \frac{3H^2}{8\pi G},$$

The density parameter (useful for comparing different cosmological models) is then defined as:

$$\Omega \equiv \frac{\rho}{\rho_c} = \frac{8\pi G\rho}{3H^2}.$$

This term originally was used as a means to determine the spatial geometry of the universe, where ρ_c is the critical density for which the spatial geometry is flat (or Euclidean). Assuming a zero vacuum energy density, if Ω is larger than unity, the space sections of the universe are closed; the universe will eventually stop expanding, then collapse. If Ω is less than unity, they are open; and the universe expands forever. However, one can also subsume the spatial curvature and vacuum energy terms into a more general expression for Ω in which case this density parameter equals exactly unity. Then it is a matter of measuring the different components, usually designated by subscripts. According to the ΛCDM model, there are important components of Ω due to baryons, cold dark matter and dark energy. The spatial geometry of the universe has been measured by the WMAP spacecraft to be nearly flat. This means that the universe can be well approximated by a model where the spatial curvature parameter k is zero; however, this does not necessarily imply that the universe is infinite: it might merely be that the universe is much larger than the part we see. (Similarly, the fact that Earth is approximately flat at the scale of the Netherlands does not imply that the Earth is flat: It only implies that it is much larger than the Netherlands).

The first Friedmann equation is often seen in terms of the present values of the density parameters, that is:

$$\frac{H^2}{H_0^2} = \Omega_{0,R} a^{-4} + \Omega_{0,M} a^{-3} + \Omega_{0,k} a^{-2} + \Omega_{0,\Lambda}.$$

Here $\Omega_{0,R}$ is the radiation density today (i.e. when $a=1$), $\Omega_{0,M}$ is the matter (dark plus baryonic) density today, $\Omega_{0,k} = 1 - \Omega_0$ is the "spatial curvature density" today, and $\Omega_{0,\Lambda}$ is the cosmological constant or vacuum density today.

Useful Solutions

The Friedmann equations can be solved exactly in presence of a perfect fluid with equation of state:

$$p = w\rho c^2,$$

where p is the pressure, ρ is the mass density of the fluid in the comoving frame and w is some constant.

In spatially flat case (k = 0), the solution for the scale factor is:

$$a(t) = a_0 t^{\frac{2}{3(w+1)}}$$

where a_0 is some integration constant to be fixed by the choice of initial conditions.

This family of solutions labelled by w is extremely important for cosmology. E.g. $w=0$ describes a matter-dominated universe, where the pressure is negligible with respect to the mass density. From the generic solution one easily sees that in a matter-dominated universe the scale factor goes as:

$$a(t) \propto t^{2/3} \text{ matter-dominated.}$$

Another important example is the case of a radiation-dominated universe, i.e., when $w = 1/3$. This leads to:

$$a(t) \propto t^{1/2} \text{ radiation dominated.}$$

Note that this solution is not valid for domination of the cosmological constant, which corresponds to an $w = -1$. In this case the energy density is constant and the scale factor grows exponentially.

Mixtures

If the matter is a mixture of two or more non-interacting fluids each with such an equation of state, then:

$$\dot{\rho}_f = -3H\left(\rho_f + \frac{p_f}{c^2}\right)$$

holds separately for each such fluid *f*. In each case,

$$\dot{\rho}_f = -3H\left(\rho_f + w_f \rho_f\right)$$

from which we get:

$$\rho_f \propto a^{-3(1+w_f)}.$$

For example, one can form a linear combination of such terms:

$$\rho = Aa^{-3} + Ba^{-4} + Ca^{0}$$

where: *A* is the density of "dust" (ordinary matter, w = 0) when $a=1$; *B* is the density of radiation (w = 1/3) when $a=1$; and *C* is the density of "dark energy" (w=−1). One then substitutes this into:

$$\left(\frac{\dot{a}}{a}\right)^2 = \frac{8\pi G}{3}\rho - \frac{kc^2}{a^2}$$

and solves for a as a function of time.

Rescaled Friedmann Equation

Set $\tilde{a} = \frac{a}{a_0}, \rho_c = \frac{3H_0^2}{8\pi G}, \Omega = \frac{\rho}{\rho_c}, t = \frac{\tilde{t}}{H_0}, \Omega_c = -\frac{kc^2}{H_0^2 a_0^2}$,

where a_0 and H_0 are separately the scale factor and the Hubble parameter today. Then we can have:

$$\frac{1}{2}\left(\frac{d\tilde{a}}{d\tilde{t}}\right)^2 + U_{\text{eff}}(\tilde{a}) = \frac{1}{2}\Omega_c$$

where $U_{\text{eff}}(\tilde{a}) = \frac{-\Omega\tilde{a}^2}{2}$. For any form of the effective potential $U_{\text{eff}}(\tilde{a})$, there is an equation of state $p = p(\rho)$ that will produce it.

Inhomogeneous Cosmology

An inhomogeneous cosmology is a physical cosmological theory (an astronomical model of the physical universe's origin and evolution) which, unlike the currently widely accepted cosmological concordance model, assumes that inhomogeneities in the distribution of matter across the universe affect local gravitational forces (i.e., at the galactic level) enough to skew our view of the Universe. When the universe began, matter was distributed homogeneously, but over billions of years, galaxies, clusters of galaxies, and superclusters have coalesced, and must, according to Einstein's theory of general relativity, warp the space-time around them. While the concordance model acknowledges this fact, it assumes that such inhomogeneities are not sufficient to affect large-scale averages of gravity in our observations. When two separate studies claimed in 1998-1999 that high redshift supernovae were further away than our calculations showed they should be, it was suggested that the expansion of the universe is accelerating, and dark energy, a repulsive energy inherent in space, was proposed to explain the acceleration. While still unexplained in nature and origin, dark energy is often claimed to comprise almost 70% of the Universe's energy density. More up to date studies have called into doubt the statistical significance of these claims, and it has been concluded that the bulk flow of the local Universe has been misrepresented as dark energy in studies of supernovae.

Inhomogeneous cosmologies assume that the backreactions of denser structures, as well as those of very empty voids, on space-time are significant enough that when not taken into account, they distort our understanding of time and our observations of distant objects. Following Thomas Buchert's publication of equations in 1997 and 2000 that derive from general relativity but also allow for the inclusion of local gravitational variations, a number of cosmological models were proposed under which the acceleration of the universe is in fact a misinterpretation of our astronomical observations and in which dark energy is unnecessary to explain them. For example, in 2007, David Wiltshire proposed a model (timescape cosmology) in which backreactions have caused time to run more slowly or, in voids, more quickly, thus giving the supernovae observed in 1998 the illusion of being further away than they were. Timescape cosmology may also imply that the expansion of the universe is in fact slowing.

Standard Cosmological Model

The conflict between the two cosmologies derives from the inflexibility of Einstein's theory of general relativity, which shows how gravity is formed by the interaction of matter, space, and time. Physicist John Wheeler famously summed up the theory's essence as "Matter tells space how to curve; space tells matter how to move." However, in order to build a workable cosmological model, all of the terms on both sides of Einstein's equations must be balanced: on one side, matter (i.e., all the things that warp time and space); on the other, the curvature of the universe and the speed at which space-time is expanding. In short, a model requires a particular amount of matter in order to produce particular curvatures and expansion rates.

In terms of matter, all modern cosmologies are founded on the cosmological principle, which states that whichever direction we look from earth, the universe is basically the same: homogeneous and isotropic (uniform in all dimensions). This principle grew out of Copernicus's assertion that there were no special observers in the universe and nothing special about the earth's location in the universe (i.e., Earth was not the center of the universe, as previously thought). Since the publication of general relativity in 1905, this homogeneity and isotropy have greatly simplified the process of devising cosmological models.

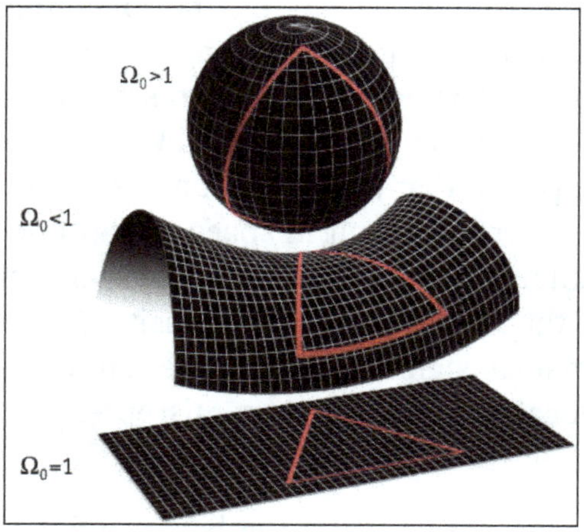

Possible shapes of the universe.

In terms of the curvature of space-time and the shape of the universe, it can theoretically be closed (positive curvature, or space-time folding in itself as though on a four-dimensional sphere's surface), open (negative curvature, with space-time folding outward), or flat (zero curvature, like the surface of a "flat" four-dimensional piece of paper).

The first real difficulty came with regards to expansion, for in 1905, as previously, the universe was assumed to be static, neither expanding nor contracting. All of Einstein's solutions to his equations in general relativity, however, predicted a dynamic universe. Therefore, in order to make his equations consistent with the apparently static universe, he added a cosmological constant, a term representing some unexplained extra energy. But when in the late 1920s Georges Lemaître's and Edwin Hubble's observations proved Alexander Friedmann's notion (derived from general relativity) that the universe was expanding, the cosmological constant became unnecessary, Einstein calling it "my greatest blunder."

With this term gone from the equation, others derived the Friedmann-Lamaître–Robertson–Walker (FLRW) solution to describe such an expanding universe, a solution built on the assumption a flat, isotropic, homogeneous universe. The FLRW model became the foundation of the standard model of a universe created by the Big Bang, and further observational evidence has helped to refine it. For example, a smooth, mostly homogeneous, and (at least when it was almost 400,000 years old) flat universe seemed to be confirmed by data from the cosmic microwave background (CMB). And after galaxies and clusters of galaxies were found in the 1970s to be rotating faster than they should without flying apart, the existence of dark matter seemed also proven, confirming its inference by Jacobus Kapteyn, Jan Oort, and Fritz Zwicky in the 1920s and 1930s and demonstrating the flexibility of the standard model. Dark matter is believed to make up roughly 23% of the energy density of the universe.

Dark Energy

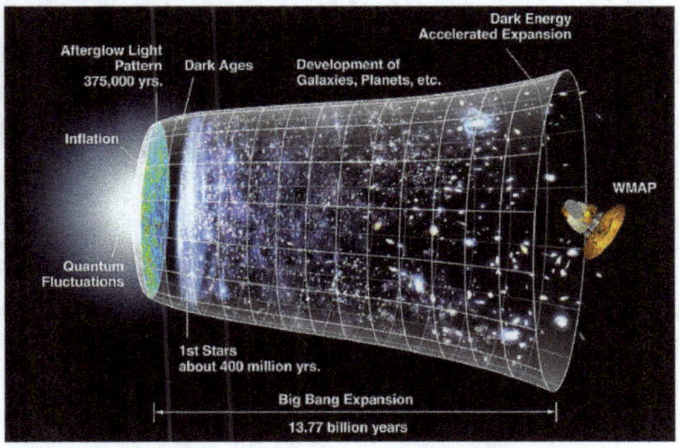

Timeline of the universe according to the CMB.

Another observation in 1998 seemed to complicate the situation further: two separate studies found distant supernovae to be fainter than expected in a steadily expanding universe; that is, they were not merely moving away from the earth but accelerating. The universe's expansion was calculated to have been accelerating since approximately 5 billion years ago. Given the gravitation braking effect that all the matter of the universe should have had on this expansion, a variation of Einstein's cosmological constant was reintroduced to represent an energy inherent in space, balancing the equations for a flat, accelerating universe. It also gave Einstein's cosmological constant new meaning, for by reintroducing it into the equation to represent dark energy, a flat universe expanding ever faster can be reproduced.

Although the nature of this energy has yet to be adequately explained, in the concordance almost 70% of the energy density of the universe in the concordance model. And thus, when including dark matter, almost 95% of the universe's energy density is explained in the concordance model by phenomena that have been inferred but not entirely explained nor directly observed. Most cosmologists still accept the concordance model, although science journalist Anil Ananthaswamy calls this agreement a "wobbly orthodoxy."

While the universe began with homogeneously distributed matter, enormous structures have since coalesced over billions of years: hundreds of billions of stars inside of galaxies, clusters of galaxies, superclusters, and vast filaments of matter. These denser regions and the voids between them must, under general relativity, have some effect, as matter dictates how space-time curves. So the extra mass of galaxies and galaxy clusters (and dark matter, should particles of ever it be directly detected) must cause nearby space-time to curve more positively, and voids should have the opposite effect, causing space-time around them to take on negative curvatures. The question is whether these effects, called backreactions, are negligible or together comprise enough to change the universe's geometry. Most scientists have assumed that they are negligible, but this has partly been because there has been no way to average space-time geometry in Einstein's equations.

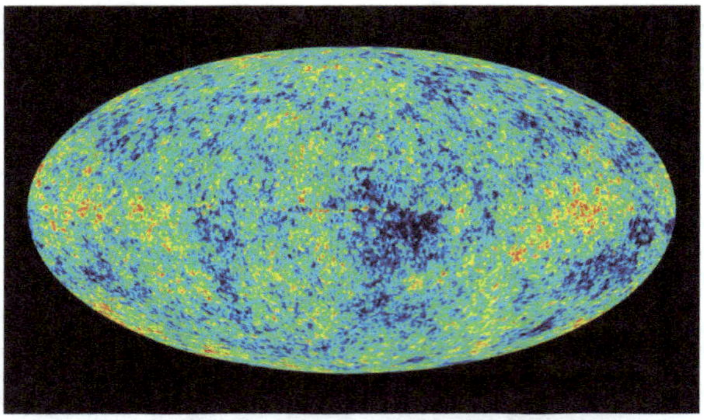

All-sky mollweide map of the CMB, created from 9 years of WMAP data. Tiny residual variations are visible, but they show a very specific pattern consistent with a hot gas that is mostly uniformly distributed.

In 2000, a set of new equations—now referred to as the set of Buchert equations—based on general relativity was published by cosmologist Thomas Buchert of the École Normale Supérieure in Lyon, France, which allow the effects of a non-uniform distribution of matter to be taken into account but still allow the behavior of the universe to be averaged. Thus, models based on a lumpy, inhomogeneous distribution of matter could now be devised. "There is no dark energy, as far as I'm concerned," Buchert told *New Scientist* in 2016. "In ten years' time, dark energy is gone." In the same article, cosmologist Syksy Räsänen said, "It's not been established beyond reasonable doubt that dark energy exists. But I'd never say that it has been established that dark energy does not exist." He also told the magazine that the question of whether backreactions are negligible in cosmology "has not been satisfactorily answered."

Inhomogeneous Cosmology

Inhomogeneous cosmology in the most general sense (assuming a totally inhomogeneous universe) is modeling the universe as a whole with the spacetime which does not possess any spacetime symmetries. Typically considered cosmological spacetimes have either the maximal symmetry, which comprises three translational symmetries and three rotational symmetries (homogeneity and isotropy with respect to every point of spacetime), the translational symmetry only (homogeneous models), or the rotational symmetry only (spherically symmetric models). Models with less symmetries (e.g. axisymmetric) are also considered as symmetric. However, it is common to call spherically symmetric models or non-homogeneous models as inhomogeneous. In inhomogeneous cosmology, the large-scale structure of the universe is modeled by exact solutions of the Einstein field equations (i.e. non-perturbatively), unlike cosmological perturbation theory, which is study of the universe that takes structure formation (galaxies, galaxy clusters, the cosmic web) into account but in a perturbative way.

Inhomogeneous cosmology usually includes the study of structure in the Universe by means of exact solutions of Einstein's field equations (i.e. *metrics*) or by spatial or spacetime averaging methods. Such models are not homogeneous, but may allow effects which can be interpreted as dark energy, or can lead to cosmological structures such as voids or galaxy clusters.

Perturbative Approach

Perturbation theory, which deals with small perturbations from e.g. a homogeneous metric, only holds as long as the perturbations are not too large, and N-body simulations use Newtonian gravity which is only a good approximation when speeds are low and gravitational fields are weak.

Non-perturbative Approach

Work towards a non-perturbative approach includes the Relativistic Zel'dovich Approximation. As of 2016, Thomas Buchert, George Ellis, Edward Kolb, and their colleagues judged that if the universe is described by cosmic variables in a backreaction scheme that includes coarse-graining and averaging, then whether dark energy is an artifact of the traditional way of using the Einstein equation remains an unanswered question.

Exact Solutions

The first historical examples of inhomogeneous (though spherically symmetric) solutions are the Lemaître–Tolman metric (or LTB model - Lemaître–Tolman-Bondi). The

Stephani metric can be spherically symmetric or totally inhomogeneous. Other examples are the Szekeres metric, Szafron metric, Barnes metric, Kustaanheimo-Qvist metric, and Senovilla metric. The Bianchi metrics as given in the Bianchi classification and Kantowski-Sachs metrics are homogeneous.

Averaging Methods

The best-known averaging approach is the scalar averaging approach, leading to the kinematical backreaction and mean 3-Ricci curvature functionals. Buchert's equations are the main equations of such averaging methods.

Timescape Cosmology

In 2007, David Wiltshire, a professor of theoretical physics at the University of Canterbury in New Zealand, argued in the *New Journal of Physics* that quasilocal variations in gravitational energy had in 1998 given the false conclusion that the expansion of the universe is accelerating. Moreover, due to the equivalence principle, which holds that gravitational and inertial energy are equivalent and thus prevents aspects of gravitational energy from being differentiated at a local level, scientists thus misidentified these aspects as dark energy. This misidentification was the result of presuming an essentially homogeneous universe, as the standard cosmological model does, and not accounting for temporal differences between matter-dense areas and voids. Wiltshire and others argued that if the universe is not only assumed not to be homogeneous but also not flat, models could be devised in which the apparent acceleration of the universe's expansion could be explained otherwise.

One more important step being left out of the standard model, Wiltshire claimed, was the fact that as proven by observation, gravity slows time. Thus, a clock will move faster in empty space, which possesses low gravitation, than inside a galaxy, which has much more gravity, and he argued that as large as a 38% difference between the time on clocks in the Milky Way and those in a galaxy floating in a void exists. Thus, unless we can correct for that—timescapes each with different times—our observations of the expansion of space will be, and are, incorrect. Wiltshire claims that the 1998 supernovae observations that led to the conclusion of an expanding universe and dark energy can instead be explained by Buchert's equations if certain strange aspects of general relativity are taken into account.

Accelerating Expansion of the Universe

The accelerating expansion of the universe is the observation that the expansion of the universe is such that the velocity at which a distant galaxy is receding from the observer is continuously increasing with time.

The accelerated expansion was discovered during 1998, by two independent projects, the Supernova Cosmology Project and the High-Z Supernova Search Team, which both used distant type Ia supernovae to measure the acceleration. The idea was that as type 1a supernovae have almost the same intrinsic brightness (a standard candle), and since objects that are further away appear dimmer, we can use the observed brightness of these supernovae to measure the distance to them. The distance can then be compared to the supernovae's cosmological redshift, which measures how much the universe has expanded since the supernova occurred. The unexpected result was that objects in the universe are moving away from one another at an accelerated rate. Cosmologists at the time expected that recession velocity would always be decelerating, due to the gravitational attraction of the matter in the universe. Three members of these two groups have subsequently been awarded Nobel Prizes for their discovery. Confirmatory evidence has been found in baryon acoustic oscillations, and in analyses of the clustering of galaxies.

The accelerated expansion of the universe is thought to have begun since the universe entered its dark-energy-dominated era roughly 4 billion years ago. Within the framework of general relativity, an accelerated expansion can be accounted for by a positive value of the cosmological constant Λ, equivalent to the presence of a positive vacuum energy, dubbed "dark energy". While there are alternative possible explanations, the description assuming dark energy (positive Λ) is used in the current standard model of cosmology, which alsoincludes cold dark matter (CDM) and is known as the Lambda-CDM model.

In the decades since the detection of cosmic microwave background (CMB) in 1965, the Big Bang model has become the most accepted model explaining the evolution of our universe. The Friedmann equation defines how the energy in the universe drives its expansion.

$$H^2 = \left(\frac{\dot{a}}{a}\right)^2 = \frac{8\pi G}{3}\rho - \frac{\kappa c^2}{a^2}$$

where κ represents the curvature of the universe, a(t) is the scale factor, ρ is the total energy density of the universe, and H is the Hubble parameter.

We define a critical density:

$$\rho_c = \frac{3H^2}{8\pi G}$$

and the density parameter,

$$\Omega = \frac{\rho}{\rho_c}$$

We can then rewrite the Hubble parameter as,

$$H(a) = H_0\sqrt{\Omega_k a^{-2} + \Omega_m a^{-3} + \Omega_r a^{-4} + \Omega_{DE} a^{-3(1+w)}}$$

where the four currently hypothesized contributors to the energy density of the universe are curvature, matter, radiation and dark energy. Each of the components decreases with the expansion of the universe (increasing scale factor), except perhaps the dark energy term. It is the values of these cosmological parameters which physicists use to determine the acceleration of the universe.

The acceleration equation describes the evolution of the scale factor with time:

$$\frac{\ddot{a}}{a} = -\frac{4\pi G}{3}\left(\rho + \frac{3P}{c^2}\right)$$

where the pressure P is defined by the cosmological model chosen.

Physicists at one time were so assured of the deceleration of the universe's expansion that they introduced a so-called deceleration parameter q_0. Current observations indicate this deceleration parameter being negative.

Relation to Inflation

According to the theory of cosmic inflation, the very early universe underwent a period of very rapid, quasi-exponential expansion. While the time-scale for this period of expansion was far shorter than that of the current expansion, this was a period of accelerated expansion with some similarities to the current epoch.

The definition of "accelerating expansion" is that the second time derivative of the cosmic scale factor, $\ddot{a}$, is positive, which is equivalent to the deceleration parameter, q, being negative. However, note this does not imply that the Hubble parameter is increasing with time. Since the Hubble parameter is defined as $H(t) \equiv \dot{a}(t) / a(t)$, it follows from the definitions that the derivative of the Hubble parameter is given by:

$$\frac{dH}{dt} = -H^2(1+q)$$

so the Hubble parameter is decreasing with time unless $q < -1$. Observations prefer $q \approx -0.55$, which implies that $\ddot{a}$ is positive but dH / dt is negative. Essentially, this implies that the cosmic recession velocity of any one particular galaxy is increasing with time, but its velocity/distance ratio is still decreasing; thus different galaxies expanding across a sphere of fixed radius cross the sphere more slowly at later times.

It is seen from above that the case of "zero acceleration/deceleration" corresponds to a(t) is a linear function of t, $q = 0$, $\dot{a} = const$, and $H(t) = 1/t$.

Evidence for Acceleration

To learn about the rate of expansion of the universe we look at the magnitude-redshift relationship of astronomical objects using standard candles, or their distance-redshift relationship using standard rulers. We can also look at the growth of large-scale structure, and find that the observed values of the cosmological parameters are best described by models which include an accelerating expansion.

Supernova Observation

The first evidence for acceleration came from the observation of Type Ia supernovae, which are exploding white dwarfs that have exceeded their stability limit. Because they all have similar masses, their intrinsic luminosity is standardizable. Repeated imaging of selected areas of the sky is used to discover the supernovae, then follow-up observations give their peak brightness, which is converted into a quantity known as luminosity distance. Spectral lines of their light can be used to determine their redshift.

Artist's impression of a Type Ia supernova, as revealed by spectro-polarimetry observations.

For supernovae at redshift less than around 0.1, or light travel time less than 10 percent of the age of the universe, this gives a nearly linear distance–redshift relation due to Hubble's law. At larger distances, since the expansion rate of the universe has changed over time, the distance-redshift relation deviates from linearity, and this deviation depends on how the expansion rate has changed over time. The full calculation requires computer integration of the Friedmann equation, but a simple derivation can be given as follows: the redshift z directly gives the cosmic scale factor at the time the supernova exploded.

$$a(t) = \frac{1}{1+z}$$

So a supernova with a measured redshift $z = 0.5$ implies the universe was $1/1 + 0.5 = 2/3$ of its present size when the supernova exploded. In the case of accelerated expansion, $\ddot{a}$ is positive therefore $\dot{a}$ was smaller in the past than today. Thus an accelerating universe took a longer time to expand from 2/3 to 1 times its present size, compared to a non-accelerating universe with constant $\dot{a}$ and the same present-day value of the Hubble constant. This results in a larger light-travel time, larger distance and fainter supernovae, which corresponds to the actual observations. Adam Riess et al. found that "the distances of the high-redshift SNe Ia were, on average, 10% to 15% farther than expected in a low mass density $\Omega_M = 0.2$ universe without a cosmological constant". This means that the measured high-redshift distances were too large, compared to nearby ones, for a decelerating universe.

Baryon Acoustic Oscillations

In the early universe before recombination and decoupling took place, photons and matter existed in a primordial plasma. Points of higher density in the photon-baryon plasma would contract, being compressed by gravity until the pressure became too large and they expanded again. This contraction and expansion created vibrations in the plasma analogous to sound waves. Since dark matter only interacts gravitationally it stayed at the centre of the sound wave, the origin of the original overdensity. When decoupling occurred, approximately 380,000 years after the Big Bang, photons separated from matter and were able to stream freely through the universe, creating the cosmic microwave background as we know it. This left shells of baryonic matter at a fixed radius from the overdensities of dark matter, a distance known as the sound horizon. As time passed and the universe expanded, it was at these anisotropies of matter density where galaxies started to form. So by looking at the distances at which galaxies at different redshifts tend to cluster, it is possible to determine a standard angular diameter distance and use that to compare to the distances predicted by different cosmological models.

Peaks have been found in the correlation function (the probability that two galaxies will be a certain distance apart) at 100 h^{-1} Mpc, indicating that this is the size of the sound horizon today, and by comparing this to the sound horizon at the time of decoupling (using the CMB), we can confirm the accelerated expansion of the universe.

Clusters of Galaxies

Measuring the mass functions of galaxy clusters, which describe the number density of the clusters above a threshold mass, also provides evidence for dark energy. By comparing these mass functions at high and low redshifts to those predicted by different cosmological models, values for w and Ω_m are obtained which confirm a low matter density and a non zero amount of dark energy.

Age of the Universe

Given a cosmological model with certain values of the cosmological density parameters, it is possible to integrate the Friedmann equations and derive the age of the universe.

$$t_0 = \int_0^1 \frac{da}{\dot{a}}$$

By comparing this to actual measured values of the cosmological parameters, we can confirm the validity of a model which is accelerating now, and had a slower expansion in the past.

Gravitational Waves as Standard Sirens

Recent discoveries of gravitational waves through LIGO and VIRGO not only confirmed Einstein's predictions but also opened a new window into the universe. These gravitational waves can work as sort of standard sirens to measure the expansion rate of the universe. Abbot et al. 2017 measured the Hubble constant value to be approximately 70 kilometres per second per megaparsec. The amplitudes of the strain 'h' is dependent on the masses of the objects causing waves, distances from observation point and gravitational waves detection frequencies. The associated distance measures are dependent on the cosmological parameters like the Hubble Constant for nearby objects and will be dependent on other cosmological parameters like the dark energy density, matter density, etc. for distant sources.

Explanatory Models

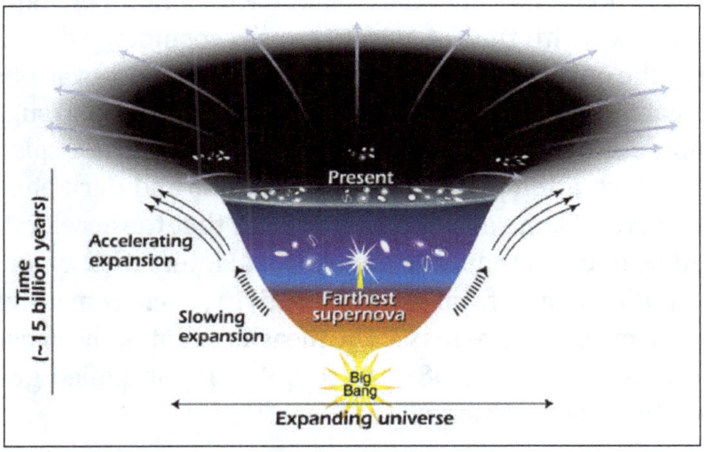

The expansion of the Universe accelerating. Time flows from bottom to top.

Dark Energy

The most important property of dark energy is that it has negative pressure (repulsive action) which is distributed relatively homogeneously in space.

$$P = wc^2\rho$$

where c is the speed of light and ρ is the energy density. Different theories of dark energy suggest different values of w, with $w < -1/3$ for cosmic acceleration (this leads to a positive value of ä in the acceleration equation above).

The simplest explanation for dark energy is that it is a cosmological constant or vacuum energy; in this case w = −1. This leads to the Lambda-CDM model, which has generally been known as the Standard Model of Cosmology from 2003 through the present, since it is the simplest model in good agreement with a variety of recent observations. Riess et al. found that their results from supernovae observations favoured expanding models with positive cosmological constant ($\Omega_\lambda > 0$) and a current accelerated expansion ($q_0 < 0$).

Phantom Energy

Current observations allow the possibility of a cosmological model containing a dark energy component with equation of state $w < -1$. This phantom energy density would become infinite in finite time, causing such a huge gravitational repulsion that the universe would lose all structure and end in a Big Rip. For example, for $w = -\frac{3}{2}$ and $H_0 = 70\ \text{km}\cdot\text{s}^{-1}\cdot\text{Mpc}^{-1}$, the time remaining before the universe ends in this Big Rip is 22 billion years.

Alternative Theories

There are many alternative explanations for the accelerating universe. Some examples are quintessence, a proposed form of dark energy with a non-constant state equation, whose density decreases with time. A negative mass cosmology does not assume that the mass density of the universe is positive (as is done in supernovae observations), and instead finds a negative cosmological constant. Occam's razor also suggests that this is the 'more parsimonious hypothesis'. Dark fluid is an alternative explanation for accelerating expansion which attempts to unite dark matter and dark energy into a single framework. Alternatively, some authors have argued that the accelerated expansion of the universe could be due to a repulsive gravitational interaction of antimatter or a deviation of the gravitational laws from general relativity, such as massive gravity, meaning that gravitons themselves have mass. The measurement of the speed of gravity with the gravitational wave event GW170817 ruled out many modified gravity theories as alternative explanation to dark energy.

Another type of model, the backreaction conjecture, was proposed by cosmologist Syksy Räsänen: the rate of expansion is not homogenous, but we are in a region where expansion is faster than the background. Inhomogeneities in the early universe cause the formation of walls and bubbles, where the inside of a bubble has less matter than on average. According to general relativity, space is less curved than on the walls, and thus appears to have more volume and a higher expansion rate. In the denser regions, the expansion is slowed by a higher gravitational attraction. Therefore, the inward collapse of the denser regions looks the same as an accelerating expansion of the bubbles, leading us to conclude that the universe is undergoing an accelerated expansion. The benefit is that it does not require any new physics such as dark energy. Räsänen does not consider the model likely, but without any falsification, it must remain a possibility. It would require rather large density fluctuations (20%) to work.

A final possibility is that dark energy is an illusion caused by some bias in measurements.

For example, if we are located in an emptier-than-average region of space, the observed cosmic expansion rate could be mistaken for a variation in time, or acceleration. A different approach uses a cosmological extension of the equivalence principle to show how space might appear to be expanding more rapidly in the voids surrounding our local cluster. While weak, such effects considered cumulatively over billions of years could become significant, creating the illusion of cosmic acceleration, and making it appear as if we live in a Hubble bubble. Yet other possibilities are that the accelerated expansion of the universe is an illusion caused by the relative motion of us to the rest of the universe, or that the supernovae sample size used wasn't large enough.

Theories for the Consequences to the Universe

As the universe expands, the density of radiation and ordinary dark matter declines more quickly than the density of dark energy and, eventually, dark energy dominates. Specifically, when the scale of the universe doubles, the density of matter is reduced by a factor of 8, but the density of dark energy is nearly unchanged (it is exactly constant if the dark energy is the cosmological constant).

In models where dark energy is the cosmological constant, the universe will expand exponentially with time in the far future, coming closer and closer to a de Sitter universe. This will eventually lead to all evidence for the Big Bang disappearing, as the cosmic microwave background is redshifted to lower intensities and longer wavelengths. Eventually, its frequency will be low enough that it will be absorbed by the interstellar medium, and so be screened from any observer within the galaxy. This will occur when the universe is less than 50 times its current age, leading to the end of cosmology as we know it as the distant universe turns dark.

A constantly expanding universe with a non-zero cosmological constant has mass density decreasing over time. In such a scenario, the current understanding is that all matter will ionize and disintegrate into isolated stable particles such as electrons and neutrinos, with all complex structures dissipating away. This scenario is known as "heat death of the universe".

Alternatives for the ultimate fate of the universe include the Big Rip mentioned above, a Big Bounce, Big Freeze or Big Crunch.

Structure Formation

In physical cosmology, structure formation is the formation of galaxies, galaxy clusters and larger structures from small early density fluctuations. The universe, as is now known from observations of the cosmic microwave background radiation, began in a hot, dense, nearly uniform state approximately 13.8 billion years ago. However, looking in the sky today, we see structures on all scales, from stars and planets to galaxies and, on still larger scales, galaxy clusters and sheet-like structures of galaxies separated by enormous voids containing few galaxies. Structure formation attempts to model how these structures formed by gravitational instability of small early density ripples.

The modern Lambda-CDM model is successful at predicting the observed large-scale distribution of galaxies, clusters and voids; but on the scale of individual galaxies there are many complications due to highly nonlinear processes involving baryonic physics, gas heating and cooling, star formation and feedback. Understanding the processes of galaxy formation is a major topic of modern cosmology research, both via observations such as the Hubble Ultra-Deep Field and via large computer simulations.

Under present models, the structure of the visible universe was formed in the following stages:

Very Early Universe

In this stage, some mechanism, such as cosmic inflation, was responsible for establishing the initial conditions of the universe: homogeneity, isotropy, and flatness. Cosmic inflation also would have amplified minute quantum fluctuations (pre-inflation) into slight density ripples of overdensity and underdensity (post-inflation).

Growth of Structure

The early universe was dominated by radiation; in this case density fluctuations larger than the cosmic horizon grow proportional to the scale factor, as the gravitational potential fluctuations remain constant. Structures smaller than the horizon remained essentially frozen due to radiation domination impeding growth. As the universe expanded, the density of radiation drops faster than matter (due to redshifting of photon energy); this led to a crossover called matter-radiation equality at ~ 50,000 years after the Big Bang. After this all dark matter ripples could grow freely, forming seeds into which the baryons could later fall. The size of the universe at this epoch forms a turnover in the matter power spectrum which can be measured in large redshift surveys.

Recombination

The universe was dominated by radiation for most of this stage, and due to the intense heat and radiation, the primordial hydrogen and helium were fully ionized into nuclei and free electrons. In this hot and dense situation, the radiation (photons) could not travel far before Thomson scattering off an electron. The universe was very hot and dense, but expanding rapidly and therefore cooling. Finally, at a little less than 400,000 years after the 'bang', it became cool enough (around 3000 K) for the protons to capture negatively charged electrons, forming neutral hydrogen atoms. (Helium atoms formed somewhat earlier due to their larger binding energy). Once nearly all the charged particles were bound in neutral atoms, the photons no longer interacted with them and were free to propagate for the next 13.8 billion years; we currently detect those photons redshifted by a factor 1090 down to 2.725 K as the Cosmic Microwave Background Radiation (CMB) filling today's universe. Several remarkable space-based missions (COBE, WMAP, Planck), have detected very slight variations in the density and temperature of the CMB. These variations were subtle, and the CMB appears very nearly uniformly the same in every direction. However, the slight temperature variations of order a few parts in 100,000 are of enormous importance, for they essentially were early "seeds" from which all subsequent complex structures in the universe ultimately developed.

The theory of what happened after the universe's first 400,000 years is one of hierarchical structure formation: The smaller gravitationally bound structures such as matter peaks containing the first stars and stellar clusters formed first, and these subsequently merged with gas and dark matter to form galaxies, followed by groups, clusters and superclusters of galaxies.

The very early universe is still a poorly understood epoch, from the viewpoint of fundamental physics. The prevailing theory, cosmic inflation, does a good job explaining the observed flatness, homogeneity and isotropy of the universe, as well as the absence of exotic relic particles (such as magnetic monopoles). Another prediction borne out by observation is that tiny perturbations in the primordial universe seed the later formation of structure. These fluctuations, while they form the foundation for all structure, appear most clearly as tiny temperature fluctuations at one part in 100,000. (To put this in perspective, the same level of fluctuations on a topographic map of the United States would show no feature taller than a few centimeters. These fluctuations are critical, because they provide the seeds from which the largest structures can grow and eventually collapse to form galaxies and stars. COBE (Cosmic Background Explorer) provided the first detection of the intrinsic fluctuations in the cosmic microwave background radiation in the 1990s.

These perturbations are thought to have a very specific character: they form a Gaussian random field whose covariance function is diagonal and nearly scale-invariant. Observed fluctuations appear to have exactly this form, and in addition the *spectral index* measured by WMAP—the spectral index measures the deviation from a scale-invariant (or Harrison-Zel'dovich) spectrum—is very nearly the value predicted by the simplest and most robust models of inflation. Another important property of the primordial perturbations, that they are adiabatic (or isentropic between the various kinds of matter that compose the universe), is predicted by cosmic inflation and has been confirmed by observations.

Other theories of the very early universe have been proposed that are claimed to make similar predictions, such as the brane gas cosmology, cyclic model, pre-big bang model and holographic universe, but they remain nascent and are not widely accepted. Some theories, such as cosmic strings, have largely been refuted by increasingly precise data.

Horizon Problem

The physical size of the Hubble radius (solid line) as a function of the scale factor of the universe. The physical wavelength of a perturbation mode (dashed line) is shown as well. The plot illustrates how the perturbation mode exits the horizon during cosmic inflation in order to reenter during radiation domination. If cosmic inflation never happened, and radiation domination continued back until a gravitational singularity, then the mode would never have exited the horizon in the very early universe.

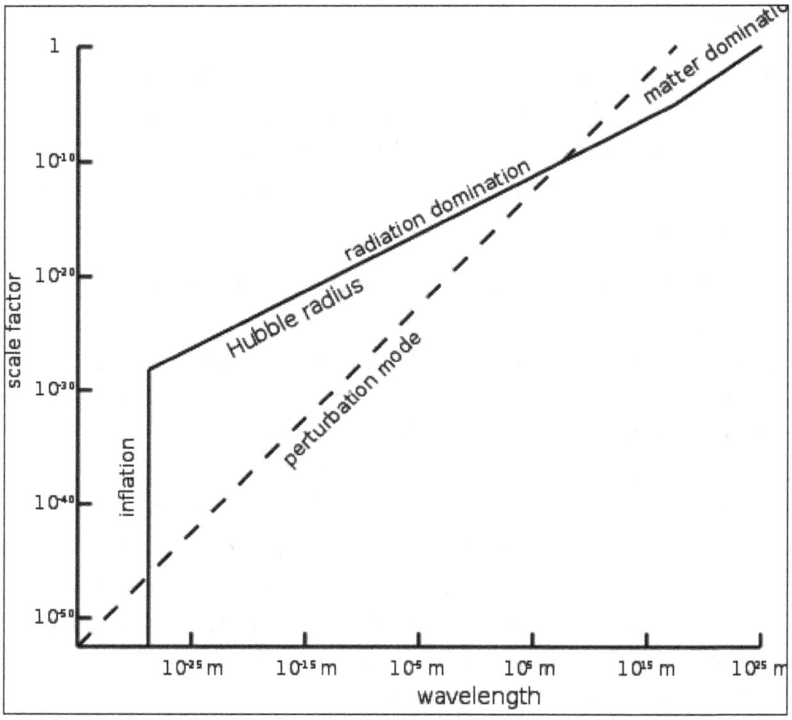

An important concept in structure formation is the notion of the Hubble radius, often called simply the *horizon,* as it is closely related to the particle horizon. The Hubble radius, which is related to the Hubble parameter H as $R = c / H$, where c is the speed of light, defines, roughly speaking, the volume of the nearby universe that has recently (in the last expansion time) been in causal contact with an observer. Since the universe is continually expanding, its energy density is continually decreasing (in the absence of truly exotic matter such as phantom energy). The Friedmann equation relates the energy density of the universe to the Hubble parameter and shows that the Hubble radius is continually increasing.

The horizon problem of big bang cosmology says that, without inflation, perturbations were never in causal contact before they entered the horizon and thus the homogeneity and isotropy of, for example, the large scale galaxy distributions cannot be explained. This is because, in an ordinary Friedmann–Lemaître–Robertson–Walker cosmology, the Hubble radius increases more rapidly than space expands, so perturbations only enter the Hubble radius, and are not pushed out by the expansion. This paradox is resolved by cosmic inflation, which suggests that during a phase of rapid expansion in the early universe the Hubble radius was nearly constant. Thus, large scale isotropy is due to quantum fluctuations produced during cosmic inflation that are pushed outside the horizon.

Primordial Plasma

The end of inflation is called reheating, when the inflation particles decay into a hot, thermal plasma of other particles. In this epoch, the energy content of the universe is entirely radiation, with standard model particles having relativistic velocities. As the plasma cools, baryogenesis and leptogenesis are thought to occur, as the quark–gluon plasma cools, electroweak symmetry breaking occurs and the universe becomes principally composed of ordinary protons, neutrons and electrons. As the universe cools further, Big Bang nucleosynthesis occurs and small quantities of deuterium, helium and lithium nuclei are created. As the universe cools and expands, the energy in photons begins to redshift away, particles become non-relativistic and ordinary matter begins to dominate the universe. Eventually, atoms begin to form as free electrons bind to nuclei. This suppresses Thomson scattering of photons. Combined with the rarefaction of the universe (and consequent increase in the mean free path of photons), this makes the universe transparent and the cosmic microwave background is emitted at recombination (the *surface of last scattering*).

Acoustic Oscillations

The primordial plasma would have had very slight overdensities of matter, thought to have derived from the enlargement of quantum fluctuations during inflation. Whatever the source, these overdensities gravitationally attract matter. But the intense heat of the near constant photon-matter interactions of this epoch rather forcefully seeks thermal equilibrium, which creates a large amount of outward pressure. These counteracting forces of gravity and pressure create oscillations, analogous to sound waves created in air by pressure differences.

These perturbations are important, as they are responsible for the subtle physics that result in the cosmic microwave background anisotropy. In this epoch, the amplitude of perturbations that enter the horizon oscillate sinusoidally, with dense regions becoming more rarefied and then becoming dense again, with a frequency which is related to the size of the perturbation. If the perturbation oscillates an integral or half-integral number of times between coming into the horizon and recombination, it appears as an acoustic peak of the cosmic microwave background anisotropy. (A half-oscillation, in which a dense region becomes a rarefied region or vice versa, appears as a peak because the anisotropy is displayed as a *power spectrum*, so underdensities contribute to the power just as much as overdensities.) The physics that determines the detailed peak structure of the microwave background is complicated, but these oscillations provide the essence.

Linear Structure

Evolution of two perturbations to the ΛCDM homogeneous big bang model. Between entering the horizon and decoupling, the dark matter perturbation (dashed line) grows logarithmically, before the growth accelerates in matter domination. On the other hand, between entering the horizon and decoupling, the perturbation in the baryon-photon fluid (solid line) oscillates rapidly. After decoupling, it grows rapidly to match the dominant matter perturbation, the dark matter mode.

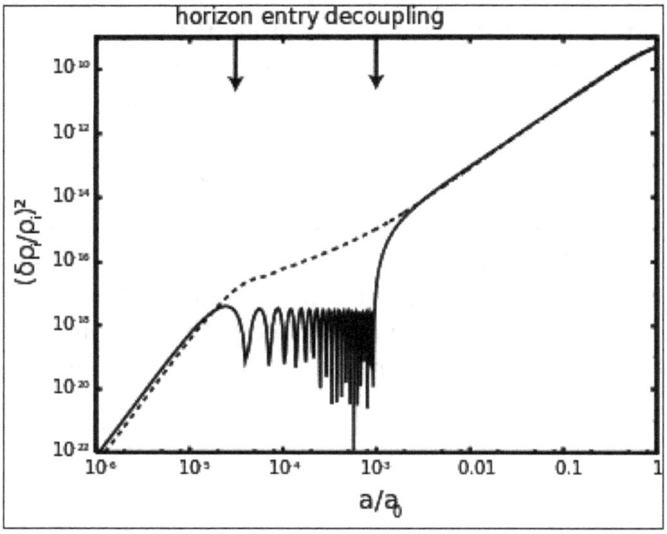

One of the key realizations made by cosmologists in the 1970s and 1980s was that the majority of the matter content of the universe was composed not of atoms, but rather a mysterious form of matter known as dark matter. Dark matter interacts through the force of gravity, but it is not composed of baryons, and it is known with very high accuracy that it does not emit or absorb radiation. It may be composed of particles that interact through the weak interaction, such as neutrinos, but it cannot be composed entirely of the three known kinds of neutrinos (although some have suggested it is a sterile neutrino). Recent evidence indicates that there are about five times as much dark matter as baryonic matter, and thus the dynamics of the universe in this epoch are dominated by dark matter.

Dark matter plays a crucial role in structure formation because it feels only the force of gravity: the gravitational Jeans instability which allows compact structures to form is not opposed by any force, such as radiation pressure. As a result, dark matter begins to collapse into a complex network of dark matter halos well before ordinary matter, which is impeded by pressure forces. Without dark matter, the epoch of galaxy formation would occur substantially later in the universe than is observed.

The physics of structure formation in this epoch is particularly simple, as dark matter perturbations with different wavelengths evolve independently. As the Hubble radius grows in the expanding universe, it encompasses larger and larger disturbances. During matter domination, all causal dark matter perturbations grow through gravitational clustering. However, the shorter-wavelength perturbations that are included during radiation domination have their growth retarded until matter domination. At this stage, luminous, baryonic matter is expected to mirror the evolution of the dark matter simply, and their distributions should closely trace one another.

It is straightforward to calculate this "linear power spectrum" and, as a tool for cosmology, it is of comparable importance to the cosmic microwave background. Galaxy surveys have measured the power spectrum, such as the Sloan Digital Sky Survey, and by surveys of the Lyman-α forest. Since these studies observe radiation emitted from galaxies and quasars, they do not directly measure the dark matter, but the large-scale distribution of galaxies (and of absorption lines in the Lyman-α forest) is expected to mirror the distribution of dark matter closely. This depends on the fact that galaxies will be larger and more numerous in denser parts of the universe, whereas they will be comparatively scarce in rarefied regions.

Nonlinear Structure

When the perturbations have grown sufficiently, a small region might become substantially denser than the mean density of the universe. At this point, the physics involved becomes substantially more complicated. When the deviations from homogeneity are small, the dark matter may be treated as a pressureless fluid and evolves by very simple equations. In regions which are significantly denser than the background, the full Newtonian theory of gravity must be included. (The Newtonian theory is appropriate because the masses involved are much less than those required to form a black hole, and the speed of gravity may be ignored as the light-crossing time for the structure is still smaller than the characteristic dynamical time.) One sign that the linear and fluid approximations become invalid is that dark matter starts to form caustics in which the trajectories of adjacent particles cross, or particles start to form orbits. These dynamics are best understood using *N*-body simulations (although a variety of semi-analytic schemes, such as the Press–Schechter formalism, can be used in some cases). While in principle these simulations are quite simple, in practice they are tough to implement, as they require simulating millions or even billions of particles. Moreover, despite the large number of particles, each particle typically weighs 10^9 solar masses and discretization effects may become significant. The largest such simulation as of 2005 is the Millennium simulation.

The result of *N*-body simulations suggests that the universe is composed largely of voids, whose densities might be as low as one-tenth the cosmological mean. The matter condenses in large filaments and haloes which have an intricate web-like structure. These form galaxy groups, clusters and superclusters. While the simulations appear to agree broadly with observations, their interpretation is complicated by the understanding of how dense accumulations of dark matter spur galaxy formation. In particular, many more small haloes form than we see in astronomical observations as dwarf galaxies and globular clusters. This is known as the galaxy bias problem, and a variety of explanations have been proposed. Most account for it as an effect in the complicated physics of galaxy formation, but some have suggested that it is a problem with our model of dark matter and that some effect, such as warm dark matter, prevents the formation of the smallest haloes.

Gas Evolution

The final stage in evolution comes when baryons condense in the centres of galaxy haloes to form galaxies, stars and quasars. Dark matter greatly accelerates the formation of dense haloes. As dark matter does not have radiation pressure, the formation of smaller structures from dark matter is impossible. This is because dark matter cannot dissipate angular momentum, whereas ordinary baryonic matter can collapse to form dense objects by dissipating angular momentum through radiative cooling. Understanding these processes is an enormously difficult computational problem, because they can involve the physics of gravity, magnetohydrodynamics, atomic physics, nuclear reactions, turbulence and even general relativity. In most cases, it is not yet possible to perform simulations that can be compared quantitatively with observations, and the best that can be achieved are approximate simulations that illustrate the main qualitative features of a process such as a star formation.

Modelling Structure Formation

Cosmological Perturbations

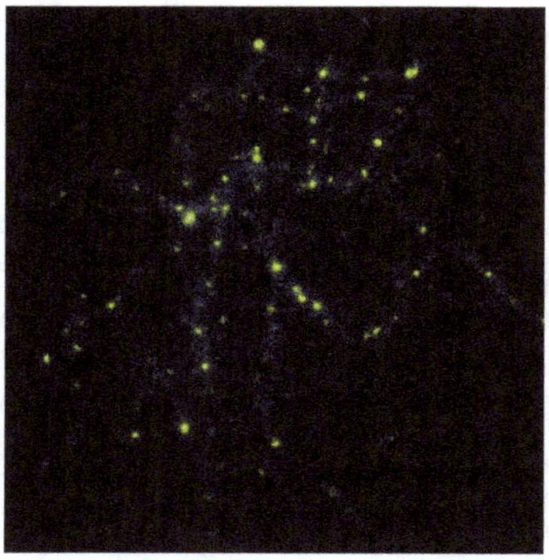

Computer simulation of large scale structure formation in a Lambda-CDM universe.

Much of the difficulty, and many of the disputes, in understanding the large-scale structure of the universe can be resolved by better understanding the choice of gauge in general relativity. By the scalar-vector-tensor decomposition, the metric includes four scalar perturbations, two vector perturbations, and one tensor perturbation. Only the scalar perturbations are significant: The vectors are exponentially suppressed in the early universe, and the tensor mode makes only a small (but important) contribution in the form of primordial gravitational radiation and the B-modes of the cosmic microwave background polarization. Two of the four scalar modes may be removed by a physically meaningless coordinate transformation. Which modes are eliminated determine the infinite number of possible gauge fixings. The most popular gauge is Newtonian gauge (and the closely related conformal Newtonian gauge), in which the retained scalars are the Newtonian potentials Φ and Ψ, which correspond exactly to the Newtonian potential energy from Newtonian gravity. Many other gauges are used, including synchronous gauge, which can be an efficient gauge for numerical computation (it is used by CMBFAST). Each gauge still includes some unphysical degrees of freedom. There is a so-called gauge-invariant formalism, in which only gauge invariant combinations of variables are considered.

Inflation and Initial Conditions

The initial conditions for the universe are thought to arise from the scale invariant quantum mechanical fluctuations of cosmic inflation. The perturbation of the background energy density at a given point $\rho(\mathbf{x},t)$ in space is then given by an isotropic, homogeneous Gaussian random field of mean zero. This means that the spatial Fourier transform of $\rho-\hat{\rho}(\mathbf{k},t)$ has the following correlation functions:

$$\langle\hat{\rho}(\mathbf{k},t)\hat{\rho}(\mathbf{k}',t)\rangle = f(k)\delta^{(3)}(\mathbf{k}-\mathbf{k}'),$$

where $\delta^{(3)}$ is the three-dimensional Dirac delta function and $k=|\mathbf{k}|$ is the length of $\mathbf{k}$. Moreover, the spectrum predicted by inflation is nearly scale invariant, which means:

$$\langle\hat{\rho}(\mathbf{k},t)\hat{\rho}(\mathbf{k}',t)\rangle = k^{n_s-1}\delta^{(3)}(\mathbf{k}-\mathbf{k}'),$$

where n_s-1 is a small number. Finally, the initial conditions are adiabatic or isentropic, which means that the fractional perturbation in the entropy of each species of particle is equal. The resulting predictions fit very well with observations, however there is a conceptual problem with the physical picture presented above. The quantum state from which the quantum fluctuations are extracted, is in fact completely homogeneous and isotropic, and thus it cannot be argued that the quantum fluctuations represent the primordial inhomogeneities and anisotropies. The interpretation of quantum uncertainties in the value of the inflation field (which is what the so-called quantum fluctuations really are) as if they were statistical fluctuations in a Gaussian random field does not follow from the application of standard rules of quantum theory. The issue is sometimes presented in terms of the "quantum to classical transition", which is a confusing manner to refer to the problem at hand, as there are very few physicists, if any, that would argue that there is any entity that is truly classical at the fundamental level. In fact, the consideration of these issues brings us face to face with the so called measurement problem in quantum theory. If anything, the problem becomes exacerbated in the cosmological context, as the early universe contains no entities that might be taken as playing the role of "observers" or of "measuring devices", both of which are essential for the standard usage of quantum mechanics. The most popular posture among cosmologists, in this regard, is to rely on arguments based on decoherence and some form of "Many Worlds Interpretation" of quantum theory. There is an intense ongoing debate about the reasonableness of that posture.

Baryogenesis

In physical cosmology, baryogenesis is the hypothetical physical process that took place during the early universe that produced baryonic asymmetry, i.e. the imbalance of matter (baryons) and antimatter (antibaryons) in the observed universe.

One of the outstanding problems in modern physics is the predominance of matter over antimatter in the universe. The universe, as a whole, seems to have a nonzero positive baryon number density – that is, matter exists. Since it is assumed in cosmology that the particles we see were created using the same physics we measure today, it would normally be expected that the overall baryon number should be zero, as matter and antimatter should have been created in equal amounts. This has led to a number of proposed mechanisms for symmetry breaking that favour the creation of normal matter (as opposed to antimatter) under certain conditions. This imbalance would have been exceptionally small, on the order of 1 in every 10000000000 (10^{10}) particles a small fraction of a second after the Big Bang, but after most of the matter and antimatter annihilated, what was left over was all the baryonic matter in the current universe, along with a much greater number of bosons. Experiments reported in 2010 at Fermilab, however, seem to show that this imbalance is much greater than previously assumed. In an experiment involving a series of particle collisions, the amount of generated matter was approximately 1% larger than the amount of generated antimatter. The reason for this discrepancy is yet unknown.

Most grand unified theories explicitly break the baryon number symmetry, which would account for this discrepancy, typically invoking reactions mediated by very massive X bosons (X) or massive Higgs bosons (H^o). The rate at which these events occur is governed largely by the mass of the intermediate X or H^o particles, so by assuming these reactions are responsible for the majority of the baryon number seen today, a maximum mass can be calculated above which the rate would be too slow to explain the presence of matter today. These estimates predict that a large volume of material will occasionally exhibit spontaneous proton decay.

Baryogenesis theories are based on different descriptions of the interaction between fundamental particles. Two main theories are electroweak baryogenesis (standard model), which would occur during the electroweak epoch, and the GUT baryogenesis, which would occur during or shortly after the grand unification epoch. Quantum field theory and statistical physics are used to describe such possible mechanisms.

Baryogenesis is followed by primordial nucleosynthesis, when atomic nuclei began to form.

The Dirac equation, formulated by Paul Dirac around 1928 as part of the development of relativistic quantum mechanics, predicts the existence of antiparticles along with the expected solutions for the corresponding particles. Since then, experiments have verified that every known kind of particle has a corresponding antiparticle. Under the CPT theorem, a particle and its antiparticle have the same mass and lifetime, and opposite charge. Given this symmetry, it is puzzling that the universe does not have equal amounts of matter and antimatter. Indeed, there is no experimental evidence that there are any significant concentrations of antimatter in the observable universe.

There are two main interpretations for this disparity: either the universe began with a small preference for matter (total baryonic number of the universe different from zero), or the universe was originally perfectly symmetric, but somehow a set of phenomena contributed to a small imbalance in favour of matter over time. The second point of view is preferred, although there is no clear experimental evidence indicating either of them to be the correct one.

Gut Baryogenesis under Sakharov Conditions

In 1967, Andrei Sakharov proposed a set of three necessary conditions that a baryon-generating interaction must satisfy to produce matter and antimatter at different rates. These conditions were inspired by the recent discoveries of the cosmic background radiation and CP-violation in the neutral kaon system. The three necessary "Sakharov conditions" are:

- Baryon number B violation.
- C-symmetry and CP-symmetry violation.
- Interactions out of thermal equilibrium.

Baryon number violation is obviously a necessary condition to produce an excess of baryons over anti-baryons. But C-symmetry violation is also needed so that the interactions which produce more baryons than anti-baryons will not be counterbalanced by interactions which produce more anti-baryons than baryons. CP-symmetry violation is similarly required because otherwise equal numbers of left-handed baryons and right-handed anti-baryons would be produced, as well as equal numbers of left-handed anti-baryons and right-handed baryons. Finally, the interactions must be out of thermal equilibrium, since otherwise CPT symmetry would assure compensation between processes increasing and decreasing the baryon number.

Currently, there is no experimental evidence of particle interactions where the conservation of baryon number is broken perturbatively: this would appear to suggest that all observed particle reactions have equal baryon number before and after. Mathematically, the commutator of the baryon number quantum operator with the (perturbative) Standard Model hamiltonian is zero: $[B,H]=BH-HB=0$. However, the Standard Model is known to violate the conservation of baryon number only non-perturbatively: a global U(1) anomaly. To account for baryon violation in baryogenesis, such events (including proton decay) can occur in Grand Unification Theories (GUTs) and supersymmetric (SUSY) models via hypothetical massive bosons such as the X boson.

The second condition – violation of CP-symmetry – was discovered in 1964 (direct CP-violation, that is violation of CP-symmetry in a decay process). Due to CPT symmetry, violation of CP-symmetry demands violation of time inversion symmetry, or T-symmetry.

In the out-of-equilibrium decay scenario, the last condition states that the rate of a reaction which generates baryon-asymmetry must be less than the rate of expansion of the universe. In this situation the particles and their corresponding antiparticles do not achieve thermal equilibrium due to rapid expansion decreasing the occurrence of pair-annihilation.

Baryogenesis within the Standard Model

Electroweak Baryogenesis

The Standard Model can incorporate baryogenesis, though the amount of net baryons (and leptons) thus created may not be sufficient to account for the present baryon asymmetry. This issue has not yet been determined decisively.

Baryogenesis within the Standard Model requires the electroweak symmetry breaking be a first-order phase transition, since otherwise sphalerons wipe off any baryon asymmetry that happened up to the phase transition, while later the amount of baryon non-conserving interactions is negligible.

The phase transition domain wall breaks the P-symmetry spontaneously, allowing for CP-symmetry violating interactions to create C-asymmetry on both its sides: quarks tend to accumulate on the broken phase side of the domain wall, while anti-quarks tend to accumulate on its unbroken phase side. This happens as follows:

Due to CP-symmetry violating electroweak interactions, some amplitudes involving quarks are not equal to the corresponding amplitudes involving anti-quarks, but rather have opposite phase since time reversal takes an amplitude to its complex conjugate, CPT-symmetry is conserved.

Though some of their amplitudes have opposite phases, both quarks and anti-quarks have positive energy, and hence acquire the same phase as they move in space-time. This phase also depends on their mass, which is identical but depends both on flavor and on the Higgs VEV which changes along the domain wall. Thus certain sums of amplitudes for quarks have different absolute values compared to those of anti-quarks. In all, quarks and anti-quarks may have different reflection and transmission probabilities through the domain wall, and it turns out that more quarks coming from the unbroken phase are transmitted compared to anti-quarks.

Thus there is a net baryonic flux through the domain wall. Due to sphaleron transitions, which are abundant in the unbroken phase, the net anti-baryonic content of the unbroken phase is wiped off. However, sphalerons are rare enough in the broken phase as not to wipe off the excess of baryons there. In total, there is net creation of baryons.

In this scenario, non-perturbative electroweak interactions (i.e. the sphaleron) are responsible for the B-violation, the perturbative electroweak Lagrangian is responsible for the CP-violation, and the domain wall is responsible for the lack of thermal equilibrium; together with the CP-violation it also creates a C-violation in each of its sides.

Matter Content in the Universe

Baryon Asymmetry Parameter

The challenges to the physics theories are then to explain *how* to produce this preference of matter over antimatter, and also the *magnitude* of this asymmetry. An important quantifier is the *asymmetry parameter*, naïvely given by:

$$\eta = \frac{n_B - n_{\bar{B}}}{n_\gamma}.$$

This quantity relates the overall number density difference between baryons and antibaryons (n_B and n_{B}, respectively) and the number density of cosmic background radiation photons n_γ.

According to the Big Bang model, matter decoupled from the cosmic background radiation (CBR) at a temperature of roughly 3000 kelvin, corresponding to an average kinetic energy of 3000 K / (10.08×10³ K/eV) = 0.3 eV. After the decoupling, the *total* number of CBR photons remains constant. Therefore, due to space-time expansion, the photon density decreases. The photon density at equilibrium temperature T per cubic centimeter, is given by:

$$n_\gamma = \frac{1}{\pi^2}\left(\frac{k_B T}{\hbar c}\right)^3 \int_0^\infty \frac{x^2}{e^x - 1} \mathrm{d}x = \frac{2\zeta(3)}{\pi^2}\left(\frac{k_B T}{\hbar c}\right)^3 \approx 20.3\left(\frac{T}{1\mathrm{K}}\right)^3 \mathrm{cm}^{-3},$$

with k_B as the Boltzmann constant, $\hbar$ as the Planck constant divided by 2π and c as the speed of light in vacuum, and $\zeta(3)$ as Apéry's constant. At the current CBR photon temperature of 2.725 K, this corresponds to a photon density n_γ of around 411 CBR photons per cubic centimeter.

Therefore, the asymmetry parameter η, as defined above, is *not* the "best" parameter. Instead, the preferred asymmetry parameter uses the entropy density s:

$$\eta_s = \frac{n_B - n_{\bar{B}}}{s}$$

because the entropy density of the universe remained reasonably constant throughout most of its evolution. The entropy density is:

$$s \stackrel{\text{def}}{=} \frac{\text{entropy}}{\text{volume}} = \frac{p+\rho}{T} = \frac{2\pi^2}{45} g_*(T) T^3$$

with p and ρ as the pressure and density from the energy density tensor $T_{\mu\nu}$, and g as the effective number of degrees of freedom for "massless" particles at temperature T (in so far as $mc^2 \ll k_B T$ holds):

$$g_*(T) = \sum_{i=\text{bosons}} g_i \left(\frac{T_i}{T}\right)^3 + \frac{7}{8} \sum_{j=\text{fermions}} g_j \left(\frac{T_j}{T}\right)^3,$$

for bosons and fermions with g_i and g_j degrees of freedom at temperatures T_i and T_j respectively. At the present epoch, $s = 7.04\, n_\gamma$.

Friedmann–Lemaître–Robertson–Walker Metric

The Friedmann–Lemaître–Robertson–Walker (FLRW) metric is an exact solution of Einstein's field equations of general relativity; it describes a homogeneous, isotropic, expanding (or otherwise, contracting) universe that is path-connected, but not necessarily simply connected. The general form of the metric follows from the geometric properties of homogeneity and isotropy; Einstein's field equations are only needed to derive the scale factor of the universe as a function of time. Depending on geographical or historical preferences, the set of the four scientists – Alexander Friedmann, Georges Lemaître, Howard P. Robertson and Arthur Geoffrey Walker – are customarily grouped as Friedmann or Friedmann–Robertson–Walker (FRW) or Robertson–Walker (RW) or Friedmann–Lemaître (FL). This model is sometimes called the Standard Model of modern cosmology, although such a description is also associated with the further developed Lambda-CDM model. The FLRW model was developed independently by the named authors in the 1920s and 1930s.

General Metric

The FLRW metric starts with the assumption of homogeneity and isotropy of space. It also assumes that the spatial component of the metric can be time-dependent. The generic metric which meets these conditions is:

$$-c^2 d\tau^2 = -c^2 dt^2 + a(t)^2 d\Sigma^2$$

where Σ ranges over a 3-dimensional space of uniform curvature, that is, elliptical space, Euclidean space, or hyperbolic space. It is normally written as a function of three spatial coordinates, but there are several conventions for doing so, detailed below. $d\Sigma$ does not depend on t — all of the time dependence is in the function a(t), known as the "scale factor".

Reduced-circumference Polar Coordinates

In reduced-circumference polar coordinates the spatial metric has the form:

$$d\Sigma^2 = \frac{dr^2}{1-kr^2} + r^2 d\Omega^2, \quad where\ d\Omega^2 = d\theta^2 + \sin^2\theta d\phi^2.$$

k is a constant representing the curvature of the space. There are two common unit conventions:

- k may be taken to have units of length^{-2}, in which case r has units of length and a(t) is unitless. k is then the Gaussian curvature of the space at the time when a(t) = 1. r is sometimes called the reduced circumference because it is equal to the measured circumference of a circle (at that value of r), centered at the origin, divided by 2π (like the r of Schwarzschild coordinates). Where appropriate, a(t) is often chosen to equal 1 in the present cosmological era, so that $d\Sigma$ measures comoving distance.

- Alternatively, k may be taken to belong to the set {−1,0,+1} (for negative, zero, and positive curvature respectively). Then r is unitless and a(t) has units of length. When k = ±1, a(t) is the radius of curvature of the space, and may also be written R(t).

A disadvantage of reduced circumference coordinates is that they cover only half of the 3-sphere in the case of positive curvature—circumferences beyond that point begin to decrease, leading to degeneracy. (This is not a problem if space is elliptical, i.e. a 3-sphere with opposite points identified).

Hyperspherical Coordinates

In hyperspherical or curvature-normalized coordinates the coordinate r is proportional to radial distance; this gives:

$$d\Sigma^2 = dr^2 + S_k(r)^2 d\Omega^2$$

where $d\Omega$ is as before and,

$$S_k(r) = \begin{cases} \sqrt{k}^{\,-1}\sin(r\sqrt{k}), & k>0 \\ r, & k=0 \\ \sqrt{|k|}^{\,-1}\sinh(r\sqrt{|k|}), & k<0. \end{cases}$$

As before, there are two common unit conventions:

- k may be taken to have units of length^{-2}, in which case r has units of length and a(t) is unitless. k is then the Gaussian curvature of the space at the time when a(t) = 1. Where appropriate, a(t) is often chosen to equal 1 in the present cosmological era, so that $d\Omega$ measures comoving distance.
- Alternatively, as before, k may be taken to belong to the set {−1,0,+1} (for negative, zero, and positive curvature respectively). Then r is unitless and a(t) has units of length. When k = ±1, a(t) is the radius of curvature of the space, and may also be written R(t). Note that, when k = +1, r is essentially a third angle along with θ and φ. The letter χ may be used instead of r.

Though it is usually defined piecewise as above, S is an analytic function of both k and r. It can also be written as a power series:

$$S_k(r) = \sum_{n=0}^{\infty} \frac{(-1)^n k^n r^{2n+1}}{(2n+1)!} = r - \frac{kr^3}{6} + \frac{k^2 r^5}{120} - \cdots$$

or as,

$$S_k(r) = r\, sinc(r\sqrt{k})$$

where sinc is the unnormalized sinc function and $\sqrt{k}$ is one of the imaginary, zero or real square roots of k. These definitions are valid for all k.

Cartesian Coordinates

When k = 0 one may write simply:

$$d\Sigma^2 = dx^2 + dy^2 + dz^2.$$

This can be extended to k ≠ 0 by defining:

$$x = r\cos\theta.$$

$$y = r\sin\theta\cos\phi.$$

$$z = r\sin\theta\sin\phi.$$

where r is one of the radial coordinates defined above, but this is rare.

Curvature

Cartesian coordinates

In flat $(k = 0)$ FLRW space using Cartesian coordinates, the surviving components of the Ricci tensor are:

$$R_{tt} = -3\frac{\ddot{a}}{a}, \quad R_{xx} = R_{yy} = R_{zz} = c^{-2}(a\ddot{a} + 2\dot{a}^2)$$

and the Ricci scalar is:

$$R = 6c^{-2}\left(\frac{\ddot{a}(t)}{a(t)} + \frac{\dot{a}^2(t)}{a^2(t)}\right).$$

Spherical Coordinates

In more general FLRW space using spherical coordinates (called "reduced-circumference polar coordinates" above), the surviving components of the Ricci tensor are:

$$R_{tt} = -3\frac{\ddot{a}}{a},$$

$$R_{rr} = \frac{c^{-2}(a(t)\ddot{a}(t) + 2\dot{a}^2(t)) + 2k}{1 - kr^2}$$

$$R_{\theta\theta} = r^2(c^{-2}(a(t)\ddot{a}(t) + 2\dot{a}^2(t)) + 2k)$$

$$R_{\phi\phi} = r^2(c^{-2}(a(t)\ddot{a}(t) + 2\dot{a}^2(t)) + 2k)\sin^2(\theta)$$

and the Ricci scalar is:

$$R = 6\left(\frac{\ddot{a}(t)}{c^2 a(t)} + \frac{\dot{a}^2(t)}{c^2 a^2(t)} + \frac{k}{a^2(t)}\right).$$

Einstein's field equations are not used in deriving the general form for the metric: It follows from the geometric properties of homogeneity and isotropy. However, determining the time evolution of $a(t)$ does require Einstein's field equations together with a way of calculating the density, $\rho(t)$, such as a cosmological equation of state.

This metric has an analytic solution to Einstein's field equations $G_{\mu\nu} + \Lambda g_{\mu\nu} = \frac{8\pi G}{c^4} T_{\mu\nu}$ giving the Friedmann equations when the energy–momentum tensor is similarly assumed to be isotropic and homogeneous. The resulting equations are:

$$\left(\frac{\dot{a}}{a}\right)^2 + \frac{kc^2}{a^2} - \frac{\Lambda c^2}{3} = \frac{8\pi G}{3}\rho$$

$$2\frac{\ddot{a}}{a} + \left(\frac{\dot{a}}{a}\right)^2 + \frac{kc^2}{a^2} - \Lambda c^2 = -\frac{8\pi G}{c^2}p.$$

These equations are the basis of the standard Big Bang cosmological model including the current ΛCDM model, and their solutions can be found here. Because the FLRW model assumes homogeneity, some popular accounts mistakenly assert that the Big Bang model cannot account for the observed lumpiness of the universe. In a strictly FLRW model, there are no clusters of galaxies, stars or people, since these are objects much denser than a typical part of the universe. Nonetheless, the FLRW model is used as a first approximation for the evolution of the real, lumpy universe because it is simple to calculate, and models which calculate the lumpiness in the universe are added onto the FLRW models as extensions. Most cosmologists agree that the observable universe is well approximated by an almost FLRW model, i.e., a model which follows the FLRW metric apart from primordial density fluctuations. As of 2003, the theoretical implications of the various extensions to the FLRW model appear to be well understood, and the goal is to make these consistent with observations from COBE and WMAP.

If the spacetime is multiply connected, then each event will be represented by more than one tuple of coordinates.

Interpretation

The pair of equations given above is equivalent to the following pair of equations;

$$\dot{\rho} = -3\frac{\dot{a}}{a}\left(\rho + \frac{p}{c^2}\right)$$

$$\frac{\ddot{a}}{a} = -\frac{4\pi G}{3}\left(\rho + \frac{3p}{c^2}\right) + \frac{\Lambda c^2}{3}$$

with k, the spatial curvature index, serving as a constant of integration for the first equation.

The first equation can be derived also from thermodynamical considerations and is equivalent to the first law of thermodynamics, assuming the expansion of the universe is an adiabatic process (which is implicitly assumed in the derivation of the Friedmann–Lemaître–Robertson–Walker metric).

The second equation states that both the energy density and the pressure cause the expansion rate of the universe $\dot{a}$ to decrease, i.e., both cause a deceleration in the expansion of the universe. This is a consequence of gravitation, with pressure playing a similar role to that of energy (or mass) density, according to the principles of general relativity. The cosmological constant, on the other hand, causes an acceleration in the expansion of the universe.

Cosmological Constant

The cosmological constant term can be omitted if we make the following replacements:

$$\rho \to \rho + \frac{\Lambda c^2}{8\pi G}$$

$$p \to p - \frac{\Lambda c^4}{8\pi G}.$$

Therefore, the cosmological constant can be interpreted as arising from a form of energy which has negative pressure, equal in magnitude to its (positive) energy density:

$$p = -\rho c^2.$$

Such form of energy—a generalization of the notion of a cosmological constant—is known as dark energy.

In fact, in order to get a term which causes an acceleration of the universe expansion, it is enough to have a scalar field which satisfies:

$$p < -\frac{\rho c^2}{3}.$$

Such a field is sometimes called quintessence.

Newtonian Interpretation

This is due to McCrea and Milne although sometimes incorrectly ascribed to Friedmann. The Friedmann equations are equivalent to this pair of equations:

$$-a^3 \dot{\rho} = 3a^2 \dot{a} \rho + \frac{3a^2 p \dot{a}}{c^2}$$

$$\frac{\dot{a}^2}{2}-\frac{G\frac{4\pi a^3}{3}\rho}{a}=-\frac{kc^2}{2}.$$

The first equation says that the decrease in the mass contained in a fixed cube (whose side is momentarily a) is the amount which leaves through the sides due to the expansion of the universe plus the mass equivalent of the work done by pressure against the material being expelled. This is the conservation of mass–energy (first law of thermodynamics) contained within a part of the universe.

The second equation says that the kinetic energy (seen from the origin) of a particle of unit mass moving with the expansion plus its (negative) gravitational potential energy (relative to the mass contained in the sphere of matter closer to the origin) is equal to a constant related to the curvature of the universe. In other words, the energy (relative to the origin) of a co-moving particle in free-fall is conserved. General relativity merely adds a connection between the spatial curvature of the universe and the energy of such a particle: positive total energy implies negative curvature and negative total energy implies positive curvature.

The cosmological constant term is assumed to be treated as dark energy and thus merged into the density and pressure terms.

During the Planck epoch, one cannot neglect quantum effects. So they may cause a deviation from the Friedmann equations.

Einstein's Radius of the Universe

Einstein's radius of the universe is the radius of curvature of space of Einstein's universe, a long-abandoned static model that was supposed to represent our universe in idealized form. Putting;

$$\dot{a}=\ddot{a}=0$$

in the Friedmann equation, the radius of curvature of space of this universe (Einstein's radius) is:

$$R_E=c/\sqrt{4\pi G\rho},$$

where c is the speed of light, G is the Newtonian gravitational constant, and ρ is the density of space of this universe. The numerical value of Einstein's radius is of the order of 10^{10} light years.

By combining the observation data from some experiments such as WMAP and Planck with theoretical results of Ehlers–Geren–Sachs theorem and its generalization, astrophysicists now agree that the universe is almost homogeneous and isotropic (when averaged over a very large scale) and thus nearly a FLRW spacetime.

Heat Death of the Universe

The heat death of the universe, also known as the Big Chill or Big Freeze, is a conjecture on the ultimate fate of the universe, which suggests the universe would evolve to a state of no thermodynamic free energy and would therefore be unable to sustain processes that increase entropy. Heat death does not imply any particular absolute temperature; it only requires that temperature differences or other processes may no longer be exploited to perform work. In the language of physics, this is when the universe reaches thermodynamic equilibrium (maximum entropy).

If the topology of the universe is open or flat, or if dark energy is a positive cosmological constant (both of which are consistent with current data), the universe will continue expanding forever, and a heat death is expected to occur, with the universe cooling to approach equilibrium at a very low temperature after a very long time period.

The hypothesis of heat death stems from the ideas of William Thomson, 1st Baron Kelvin (Lord Kelvin), who in the 1850s took the theory of heat as mechanical energy loss in nature (as embodied in the first two laws of thermodynamics) and extrapolated it to larger processes on a universal scale.

The idea of heat death stems from the second law of thermodynamics, of which one version states that entropy tends to increase in an isolated system. From this, the hypothesis implies that if the universe lasts for a sufficient time, it will asymptotically approach a state where all energy is evenly distributed. In other words, according to this hypothesis, there is a tendency in nature to the dissipation (energy transformation) of mechanical energy (motion) into thermal energy; hence, by extrapolation, there exists the view that, in time, the mechanical movement of the universe will run down as work is converted to heat because of the second law.

The conjecture that all bodies in the universe cool off, eventually becoming too cold to support life, seems to have been first put forward by the French astronomer Jean Sylvain Bailly in 1777 in his writings on the history of astronomy and in the ensuing correspondence with Voltaire. In Bailly's view, all planets have an internal heat and are now at some particular stage of cooling. Jupiter, for instance, is still too hot for life to arise there for thousands of years, while the Moon is already too cold. The final state, in this view, is described as one of "equilibrium" in which all motion ceases.

The idea of heat death as a consequence of the laws of thermodynamics, however, was first proposed in loose terms beginning in 1851 by William Thomson, who theorized further on the mechanical energy loss views of Sadi Carnot, James Joule, and Rudolf Clausius. Thomson's views were then elaborated on more definitively over the next decade by Hermann von Helmholtz and William Rankine.

Lord Kelvin originated the idea of universal heat death in 1852.

The idea of heat death of the universe derives from discussion of the application of the first two laws of thermodynamics to universal processes. Specifically, in 1851, William Thomson outlined the view, as based on recent experiments on the dynamical theory of heat: "Heat is not a substance, but a dynamical form of mechanical effect, we perceive that there must be an equivalence between mechanical work and heat, as between cause and effect."

In 1852, Thomson published *On a Universal Tendency in Nature to the Dissipation of Mechanical Energy*, in which he outlined the rudiments of the second law of thermodynamics summarized by the view that mechanical motion and the energy used to create that motion will naturally tend to dissipate or run down. The ideas in this paper, in relation to their application to the age of the Sun and the dynamics of the universal operation, attracted the likes of William Rankine and Hermann von Helmholtz. The three of them were said to have exchanged ideas on this subject. In 1862, Thomson published "On the age of the Sun's heat", an article in which he reiterated his fundamental beliefs in the indestructibility of energy (the first law) and the universal dissipation of energy (the second law), leading to diffusion of heat, cessation of useful motion (work), and exhaustion of potential energy through the material universe, while clarifying his view of the consequences for the universe as a whole. In a key paragraph, Thomson wrote:

"The result would inevitably be a state of universal rest and death, if the universe were finite and left to obey existing laws. But it is impossible to conceive a limit to the extent of matter in the universe; and therefore science points rather to an endless progress, through an endless space, of action involving the transformation of potential energy into palpable motion and hence into heat, than to a single finite mechanism, running down like a clock, and stopping for ever."

In the years to follow both Thomson's 1852 and the 1865 papers, Helmholtz and Rankine both credited Thomson with the idea, but read further into his papers by publishing views stating that Thomson argued that the universe will end in a "*heat death*" (Helmholtz) which will be the "*end of all physical phenomena*" (Rankine).

Proposals about the final state of the universe depend on the assumptions made about its ultimate fate, and these assumptions have varied considerably over the late 20th century and early 21st century. In a hypothesized "open" or "flat" universe that continues expanding indefinitely, either a heat death or a Big Rip is expected to eventually occur. If the cosmological constant is zero, the universe will approach absolute zero temperature over a very long timescale. However, if the cosmological constant is positive, as appears to be the case in recent observations, the temperature will asymptote to a non-zero positive value, and the universe will approach a state of maximum entropy in which no further work is possible.

If a Big Rip does not happen long before that, the "heat death" situation could be avoided if there is a method or mechanism to regenerate hydrogen atoms from radiation, dark matter, dark energy, zero-point energy, or other sources. If so, it is at least possible that star formation and heat transfer can continue, avoiding a gradual running down of the universe due to the conversion of matter into energy and heavier elements in stellar processes, and the absorption of matter by black holes and their subsequent evaporation as Hawking radiation.

Time Frame for Heat Death

From the Big Bang through the present day, matter and dark matter in the universe are thought to have been concentrated in stars, galaxies, and galaxy clusters, and are presumed to continue to be so well into the future. Therefore, the universe is not in thermodynamic equilibrium, and objects can do physical work. The decay time for a supermassive black hole of roughly 1 galaxy mass (10^{11} solar masses) due to Hawking radiation is on the order of 10^{100} years, so entropy can be produced until at least that time. Some large black holes in the universe are predicted to continue to grow up to perhaps 10^{14} $M_\odot$ during the collapse of superclusters of galaxies. Even these would evaporate over a timescale of up to 10^{106} years. After that time, the universe enters the so-called Dark Era and is expected to consist chiefly of a dilute gas of photons and leptons. With only very diffuse matter remaining, activity in the universe will have tailed off dramatically, with extremely low energy levels and extremely long timescales. Speculatively, it is possible that the universe may enter a second inflationary epoch, or assuming that the current vacuum state is a false vacuum, the vacuum may decay into a lower-energy state. It is also possible that entropy production will cease and the universe will reach heat death. Another universe could possibly be created by random quantum fluctuations or quantum tunneling in roughly $10^{10^{10^{56}}}$ years. Over vast periods of time, a spontaneous entropy *decrease* would eventually occur via the Poincaré recurrence theorem, thermal fluctuations, and fluctuation theorem. Such a scenario, however, has been described as "highly speculative, probably wrong, and completely untestable". Sean M. Carroll, originally an advocate of this idea, no longer supports it.

Max Planck wrote that the phrase "entropy of the universe" has no meaning because it admits of no accurate definition. More recently, Walter Grandy writes: "It is rather presumptuous to speak of the entropy of a universe about which we still understand so little, and we wonder how one might define thermodynamic entropy for a universe and its major constituents that have never been in equilibrium in their entire existence." According to Tisza: "If an isolated system is not in equilibrium, we cannot associate an entropy with it." Buchdahl writes of "the entirely unjustifiable assumption that the universe can be treated as a closed thermodynamic system". According to Gallavotti: "There is no universally accepted notion of entropy for systems out of equilibrium, even when in a stationary state." Discussing the question of entropy for non-equilibrium states in general, Lieb and Yngvason express their opinion as follows: "Despite the fact that most physicists believe in such a nonequilibrium entropy, it has so far proved impossible to define it in a clearly satisfactory way." In Landsberg's opinion: "The *third* misconception is that thermodynamics, and in particular, the concept of entropy, can without further enquiry be applied to the whole universe."

A recent analysis of entropy states, "The entropy of a general gravitational field is still not known", and, "gravitational entropy is difficult to quantify". The analysis considers several possible assumptions that would be needed for estimates and suggests that the observable universe has more entropy than previously thought. This is because the analysis concludes that supermassive black holes are the largest contributor. Lee Smolin goes further: "It has long been known that gravity is important for keeping the universe out of thermal equilibrium. Gravitationally bound systems have negative specific heat—that is, the velocities of their components increase when energy is removed. Such a system does not evolve toward a homogeneous equilibrium state. Instead it becomes increasingly structured and heterogeneous as it fragments into subsystems."

Big Rip

In physical cosmology, the Big Rip is a hypothetical cosmological model concerning the ultimate fate of the universe, in which the matter of the universe, from stars and galaxies to atoms and subatomic particles, and even spacetime itself, is progressively torn apart by the expansion of the universe at a certain time in the future. According to the standard model of cosmology the scale factor of the universe is known to be accelerating and, in the future era of cosmological constant dominance, will increase exponentially. However, this expansion is similar for every moment of time (hence the exponential law – the expansion of a local volume is the same number of times over the same time interval), and is characterized by an unchanging, small Hubble constant, effectively ignored by any bound material structures. By contrast in the Big Rip scenario the Hubble constant increases to infinity in a finite time.

The possibility of sudden rip singularity occurs only for hypothetical matter (phantom energy) with implausible physical properties.

The truth of the hypothesis relies on the type of dark energy present in our universe. The type that could prove this hypothesis is a constantly increasing form of dark energy. If the dark energy in the universe increases without limit, it could overcome all forces that hold the universe together. The key value is the equation of state parameter w, the ratio between the dark energy pressure and its energy density. If $w < -1$, this dynamical vacuum energy is known as phantom energy, an extreme form of quintessence.

Expansion

A universe dominated by phantom energy is an accelerating universe, expanding at an ever-increasing rate. However, this implies that the size of the observable universe is continually shrinking; the distance to the edge of the observable universe which is moving away at the speed of light from any point moves ever closer. When the size of the observable universe becomes smaller than any particular structure, no interaction by any of the fundamental forces can occur between the most remote parts of the structure. When these interactions become impossible, the structure is "ripped apart". The model implies that after a finite time there will be a final singularity, called the "Big Rip", in which all distances diverge to infinite values.

The authors of this hypothesis, led by Robert R. Caldwell of Dartmouth College, calculate the time from the present to the end of the universe as we know it for this form of energy to be:

$$t_{\text{rip}} - t_0 \approx \frac{2}{3|1+w|H_0\sqrt{1-\Omega_{\text{m}}}}$$

where w is defined above, H_0 is Hubble's constant and Ω_m is the present value of the density of all the matter in the universe.

A hypothetical example with $w = -1.5$, $H_0 = 70$ km/s/Mpc, and $\Omega_m = 0.3$, in which case the Big Rip would happen approximately 22 billion years from the present.

For $w = -1.5$, the galaxies would first be separated from each other. About 60 million years before the Big Rip, gravity would be too weak to hold the Milky Way and other individual galaxies together. Galaxies would be destroyed as stars separate from the main black hole. Approximately three months before the Big Rip, the Solar System (or systems similar to our own at this distant time, as the survival of our Solar System 22 billion years in the future is a questionable assumption) would be gravitationally unbound. Planets would be detached from the star's orbit. In the last minutes, stars and planets would be torn apart, and an extremely short amount of time before the Big Rip, atoms would be destroyed. At the time the Big Rip occurs, even spacetime itself would be ripped apart and the scale factor would be infinity.

Observed Universe

Evidence indicates w to be very close to -1 in our universe, which makes w the dominating term in the equation. The closer that w is to -1, the closer the denominator is to zero and the further the Big Rip is in the future. If w were exactly equal to -1, the Big Rip could not happen, regardless of the values of H_0 or Ω_m.

According to the latest cosmological data available, the uncertainties are still too large to discriminate among the three cases $w < -1$, $w = -1$, and $w > -1$.

Big Bounce

The Big Bounce is a hypothetical cosmological model for the origin of the known universe. It was originally suggested as a phase of the *cyclic model* or *oscillatory universe* interpretation of the Big Bang, where the first cosmological event was the result of the collapse of a previous universe. It receded from serious consideration in the early 1980s after inflation theory emerged as a solution to the horizon problem, which had arisen from advances in observations revealing the large-scale structure of the universe. In the early 2000s, inflation was found by some theorists to be problematic and unfalsifiable in that its various parameters could be adjusted to fit any observations, so that the properties of the observable universe are a matter of chance. Alternative pictures including a Big Bounce may provide a predictive and falsifiable possible solution to the horizon problem, and are under active investigation as of 2017.

Big bounce models have a venerable history and were endorsed on largely aesthetic grounds by cosmologists including Willem de Sitter, Carl Friedrich von Weizsäcker, George McVittie and George Gamow (who stressed that "from the physical point of view we must forget entirely about the precollapse period").

By the early 1980s, the advancing precision and scope of observational cosmology had revealed that the large-scale structure of the universe is flat, homogenous and isotropic, a finding later accepted as the Cosmological Principle to apply at scales beyond roughly 300 million light-years. It was recognized that it was necessary to find an explanation for how distant regions of the universe could have essentially identical properties without ever having been in light-like communication. A solution was proposed to be a period of exponential expansion of space in the early universe, as a basis for what became known as Inflation theory. Following the brief inflationary period, the universe continues to expand, but at a less rapid rate.

Various formulations of inflation theory and their detailed implications became the subject of intense theoretical study. In the absence of a compelling alternative, inflation became the leading solution to the horizon problem. In the early 2000s, inflation was found by some theorists to be problematic and unfalsifiable in that its various parameters could be adjusted to fit any observations, a situation known as a fine-tuning problem. Furthermore, inflation was found to be inevitably eternal, creating an infinity of different universes with typically different properties, so that the properties of the observable universe are a matter of chance. An alternative concept including a Big Bounce was conceived as a predictive and falsifiable possible solution to the horizon problem, and is under active investigation as of 2017.

The phrase "Big Bounce" appeared in the scientific literature in 1987, when it was first used in the title of a pair of articles in *Stern und Weltraum* by Wolfgang Priester and Hans-Joachim Blome. It reappeared in 1988 in Iosif Rozental's *Big Bang, Big Bounce*, a revised English-language translation of a Russian-language book, and in a 1991 article by Priester and Blome in *Astronomy and Astrophysics*. (The phrase apparently originated as the title of a novel by Elmore Leonard in 1969, shortly after increased public awareness of the Big Bang model with of the discovery of the cosmic microwave background by Penzias and Wilson in 1965.)

Martin Bojowald, an assistant professor of physics at Pennsylvania State University, published a study in July 2007 detailing work somewhat related to loop quantum gravity that claimed to mathematically solve the time before the Big Bang, which would give new weight to the oscillatory universe and Big Bounce theories.

One of the main problems with the Big Bang theory is that at the moment of the Big Bang, there is a singularity of zero volume and infinite energy. This is normally interpreted as the end of the physics as we know it; in this case, of the theory of general relativity. This is why one expects quantum effects to become important and avoid the singularity.

However, research in loop quantum cosmology purported to show that a previously existing universe collapsed, not to the point of singularity, but to a point before that where the quantum effects of gravity become so strongly repulsive that the universe rebounds back out, forming a new branch. Throughout this collapse and bounce, the evolution is unitary.

Bojowald also claims that some properties of the universe that collapsed to form ours can also be determined. Some properties of the prior universe are not determinable however due to some kind of uncertainty principle.

This work is still in its early stages and very speculative. Some extensions by further scientists have been published in *Physical Review* Letters.

In 2003, Peter Lynds has put forward a new cosmology model in which time is cyclic. In his theory our Universe will eventually stop expanding and then contract. Before becoming a singularity, as one would expect from Hawking's black hole theory, the universe would bounce. Lynds claims that a singularity would violate the second law of thermodynamics and this stops the universe from being bounded by singularities. The Big Crunch would be avoided with a new Big Bang. Lynds suggests the exact history of the universe would be repeated in each cycle in an eternal recurrence. Some critics argue that while the universe may be cyclic, the histories would all be variants. Lynds' theory has been dismissed by mainstream physicists for the lack of a mathematical model behind its philosophical considerations.

In 2006, it was proposed that the application of loop quantum gravity techniques to Big Bang cosmology can lead to a bounce that need not be cyclic.

In 2011, Nikodem Popławski showed that a nonsingular Big Bounce appears naturally in the Einstein-Cartan-Sciama-Kibble theory of gravity. This theory extends general relativity by removing a constraint of the symmetry of the affine connection and regarding its antisymmetric part, the torsion tensor, as a dynamical variable. The minimal coupling between torsion and Dirac spinors generates a spin-spin interaction which is significant in fermionic matter at extremely high densities. Such an interaction averts the unphysical Big Bang singularity, replacing it with a cusp-like bounce at a finite minimum scale factor, before which the universe was contracting. This scenario also explains why the present Universe at largest scales appears spatially flat, homogeneous and isotropic, providing a physical alternative to cosmic inflation.

In 2012, a new theory of nonsingular big bounce was successfully constructed within the frame of standard Einstein gravity. This theory combines the benefits of matter bounce and Ekpyrotic cosmology. Particularly, the famous BKL instability, that the homogeneous and isotropic background cosmological solution is unstable to the growth of anisotropic stress, is resolved in this theory. Moreover, curvature perturbations seeded in matter contraction are able to form a nearly scale-invariant primordial power spectrum and thus provides a consistent mechanism to explain the cosmic microwave background (CMB) observations.

A few sources argue that distant supermassive black holes whose large size is hard to explain so soon after the Big Bang, such as ULAS J1342+0928, may be evidence for a Big Bounce, with these supermassive black holes being formed before the Big Bounce.

Galaxy Filament

In physical cosmology, galaxy filaments (subtypes: Supercluster complexes, galaxy walls, and galaxy sheets) are the largest known structures in the universe. They are massive, thread-like formations, with a typical length of 50 to 80 megaparsecs h^{-1} (or of the order of 200 to 500 million light-years) that form the boundaries between large voids in the universe. Filaments consist of gravitationally bound galaxies. Parts wherein many galaxies are very close to one another (in cosmic terms) are called superclusters.

Formation

In the standard model of the evolution of the universe, galactic filaments form along and follow web-like strings of dark matter. It is thought that this dark matter dictates the structure of the Universe on the grandest of scales. Dark matter gravitationally attracts baryonic matter, and it is this "normal" matter that astronomers see forming long, thin walls of super-galactic clusters.

Filaments

Filament subtype of filaments have roughly similar major and minor axes in cross-section, along the lengthwise axis.

Filaments of Galaxies				
Filament	Date	Mean distance	Dimension	Notes
Coma Filament				The Coma Supercluster lies within the Coma Filament. It forms part of the CfA2 Great Wall.
Perseus–Pegasus Filament	1985			Connected to the Pisces–Cetus Supercluster, with the Perseus–Pisces Supercluster being a member of the filament.
Ursa Major Filament				Connected to the CfA Homunculus, a portion of the filament forms a portion of the "leg" of the Homunculus.
Lynx–Ursa Major Filament (LUM Filament)	1999	from 2000 km/s to 8000 km/s in redshift space		Connected to and separate from the Lynx–Ursa Major Supercluster.
z=2.38 filament around protocluster ClG J2143-4423	2004	z=2.38	110Mpc	A filament the length of the Great Wall was discovered in 2004. As of 2008, it was still the largest structure beyond redshift 2.

A short filament, detected by identifying an alignment of star-forming galaxies, in the neighborhood of the Milky Way and the Local Group was proposed by Adi Zitrin and Noah Brosch. The reality of this filament, and the identification of a similar but shorter filament, were the result of a study by McQuinn et al. based on distance measurements using the TRGB method.

Galaxy Walls

The galaxy wall subtype of filaments have a significantly greater major axis than minor axis in cross-section, along the lengthwise axis.

Walls of Galaxies				
Wall	Date	Mean distance	Dimension	Notes
CfA2 Great Wall (Coma Wall, Great Wall, Northern Great Wall, Great Northern Wall, CfA Great Wall)	1989	z=0.03058	251Mpc long 750 Mly long 250 Mly wide 20 Mly thick	This was the first super-large large-scale structure or pseudo-structure in the universe to be discovered. The CfA Homunculus lies at the heart of the Great Wall, and the Coma Supercluster forms most of the homunculus structure. The Coma Cluster lies at the core.
Sloan Great Wall (SDSS Great Wall)	2003	z=0.07804	433Mpc long	This was the largest known galaxy filament to be discovered, until it was eclipsed by the Hercules–Corona Borealis Great Wall found ten years later.
Sculptor Wall (Southern Great Wall, Great Southern Wall, Southern Wall)			8000 km/s long 5000 km/s wide 1000 km/s deep (in redshift space dimensions)	The Sculptor Wall is "parallel" to the Fornax Wall and "perpendicular" to the Grus Wall.
Grus Wall				The Grus Wall is "perpendicular" to the Fornax and Sculptor Walls.
Fornax Wall				The Fornax Cluster is part of this wall. The wall is "parallel" to the Sculptor Wall and "perpendicular" to the Grus Wall.
Hercules–Corona Borealis Great Wall	2013	z≈	3 Gpc long, 150 000 km/s deep (in redshift space)	The largest known structure in the universe. This is also the first time since 1991 that a galaxy filament/great wall held the record as the largest known structure in the universe.

- A "Centaurus Great Wall" (or "Fornax Great Wall" or "Virgo Great Wall") has been proposed, which would include the Fornax Wall as a portion of it (visually created by the Zone of Avoidance) along with the Centaurus Supercluster and the Virgo Supercluster also known as our Local Supercluster within which the Milky Way galaxy is located (implying this to be the Local Great Wall).

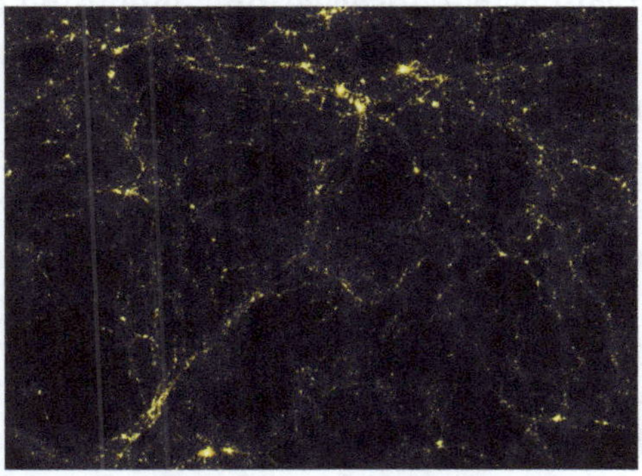

Galaxy filaments, walls and voids form web-like structures.

- A wall was proposed to be the physical embodiment of the Great Attractor, with the Norma Cluster as part of it. It is sometimes referred to as the Great Attractor Wall or Norma Wall. This suggestion was superseded by the proposal of a supercluster, Laniakea, that would encompass the Great Attractor, Virgo Supercluster, Hydra-Centaurus Superclusters.
- A wall was proposed in 2000 to lie at z=1.47 in the vicinity of radio galaxy B3 0003+387.
- A wall was proposed in 2000 to lie at z=0.559 in the northern Hubble Deep Field (HDF North).

Map of Nearest Galaxy Walls

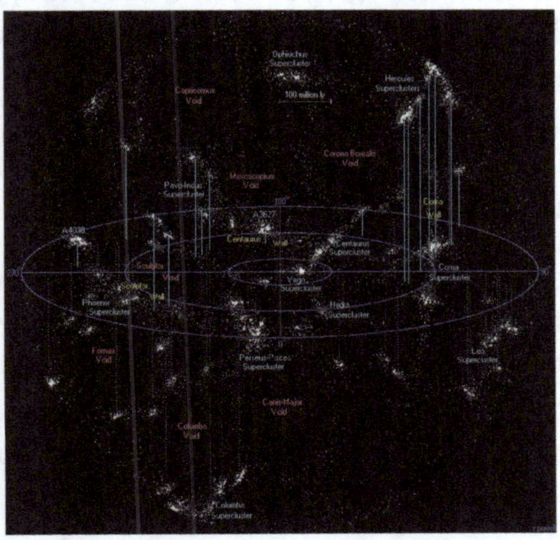

The Universe within 500 million light years, showing the nearest galaxy walls.

Dark Matter

Dark matter is a form of matter thought to account for approximately 85% of the matter in the universe and about a quarter of its total energy density. Its presence is implied in a variety of astrophysical observations, including gravitational effects that cannot be explained by accepted theories of gravity unless more matter is present than can be seen. For this reason, most experts think that dark matter is abundant in the universe and that it has had a strong influence on its structure and evolution. Dark matter is called dark because it does not appear to interact with observable electromagnetic radiation, such as light, and so it is undetectable by existing astronomical instruments.

Primary evidence for dark matter comes from calculations showing that many galaxies would fly apart, or that they would not have formed or would not move as they do, if they did not contain a large amount of unseen matter. Other lines of evidence include observations in gravitational lensing and in the cosmic microwave background, along with astronomical observations of the observable universe's current structure, the formation and evolution of galaxies, mass location during galactic collisions, and the motion of galaxies within galaxy clusters. In the standard Lambda-CDM model of cosmology, the total mass–energy of the universe contains 5% ordinary matter and energy, 27% dark matter and 68% of an unknown form of energy known as dark energy. Thus, dark matter constitutes 85% of total mass, while dark energy plus dark matter constitute 95% of total mass–energy content.

Because dark matter has not yet been observed directly, if it exists, it must barely interact with ordinary baryonic matter and radiation, except through gravity. Most dark matter is thought to be non-baryonic in nature; it may be composed of some as-yet undiscovered subatomic particles. The primary candidate for dark matter is some new kind of elementary particle that has not yet been discovered, in particular, weakly-interacting massive particles (WIMPs). Many experiments to directly detect and study dark matter particles are being actively undertaken, but none have yet succeeded. Dark matter is classified as "cold", "warm", or "hot" according to its velocity (more precisely, its free streaming length). Current models favor a cold dark matter scenario, in which structures emerge by gradual accumulation of particles.

Although the existence of dark matter is generally accepted by the scientific community, some astrophysicists, intrigued by certain observations which do not fit the dark matter theory, argue for various modifications of the standard laws of general relativity, such as modified Newtonian dynamics, tensor–vector–scalar gravity, or entropic gravity. These models attempt to account for all observations without invoking supplemental non-baryonic matter.

In standard cosmology, matter is anything whose energy density scales with the inverse cube of the scale factor, i.e., $\rho \propto a^{-3}$. This is in contrast to radiation, which scales as the inverse fourth power of the scale factor $\rho \propto a^{-4}$, and a cosmological constant, which is independent of a. These scalings can be understood intuitively: For an ordinary particle in a cubical box, doubling the length of the sides of the box decreases the density (and hence energy density) by a factor of 8 (= 2^3). For radiation, the energy density decreases by a factor of 16 (= 2^4), because any act whose effect increases the scale factor must also cause a proportional redshift. A cosmological constant, as an intrinsic property of space, has a constant energy density regardless of the volume under consideration.

In principle, "dark matter" means all components of the universe which are not visible but still obey $\rho \propto a^{-3}$. In practice, the term "dark matter" is often used to mean only the non-baryonic component of dark matter, i.e., excluding "missing baryons." Context will usually indicate which meaning is intended.

Observational Evidence

Galaxy Rotation Curves

The arms of spiral galaxies rotate around the galactic center. The luminous mass density of a spiral galaxy decreases as one goes from the center to the outskirts. If luminous mass were all the matter, then we can model the galaxy as a point mass in the centre and test masses orbiting around it, similar to the Solar System. From Kepler's Second Law, it is expected that the rotation velocities will decrease with distance from the center, similar to the Solar System. This is not observed. Instead, the galaxy rotation curve remains flat as distance from the center increases.

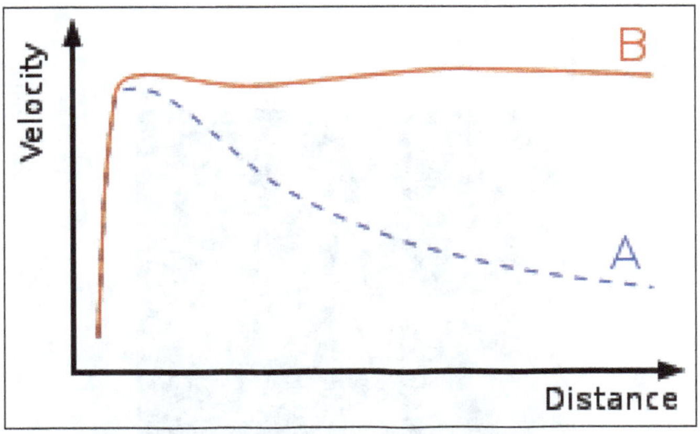

Rotation curve of a typical spiral galaxy: predicted (A) and observed (B). Dark matter can explain the 'flat' appearance of the velocity curve out to a large radius.

If Kepler's laws are correct, then the obvious way to resolve this discrepancy is to conclude the mass distribution in spiral galaxies is not similar to that of the Solar System. In particular, there is a lot of non-luminous matter (dark matter) in the outskirts of the galaxy.

Velocity Dispersions

Stars in bound systems must obey the virial theorem. The theorem, together with the measured velocity distribution, can be used to measure the mass distribution in a bound system, such as elliptical galaxies or globular clusters. With some exceptions, velocity dispersion estimates of elliptical galaxies do not match the predicted velocity dispersion from the observed mass distribution, even assuming complicated distributions of stellar orbits.

As with galaxy rotation curves, the obvious way to resolve the discrepancy is to postulate the existence of non-luminous matter.

Galaxy Clusters

Galaxy clusters are particularly important for dark matter studies since their masses can be estimated in three independent ways:

- From the scatter in radial velocities of the galaxies within clusters.
- From X-rays emitted by hot gas in the clusters. From the X-ray energy spectrum and flux, the gas temperature and density can be estimated, hence giving the pressure; assuming pressure and gravity balance determines the cluster's mass profile.
- Gravitational lensing (usually of more distant galaxies) can measure cluster masses without relying on observations of dynamics (e.g., velocity).

Generally, these three methods are in reasonable agreement dark matter outweighs visible matter by approximately 5 to 1.

Gravitational Lensing

Strong gravitational lensing as observed by the Hubble Space Telescope in Abell 1689 indicates the presence of dark matter.

One of the consequences of general relativity is massive objects (such as a cluster of galaxies) lying between a more distant source (such as a quasar) and an observer should act as a lens to bend the light from this source. The more massive an object, the more lensing is observed.

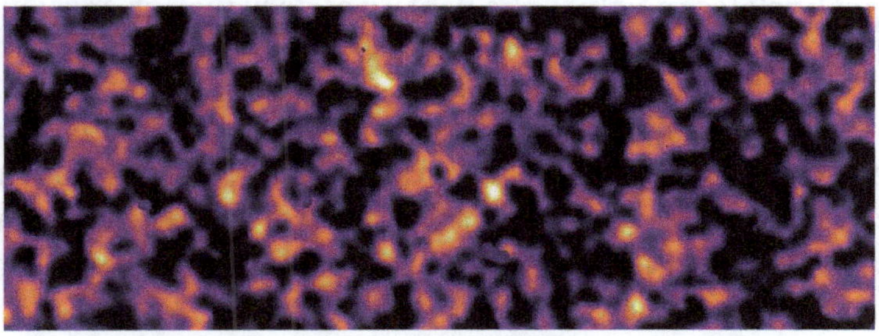

Dark matter map for a patch of sky based on gravitational lensing analysis of a Kilo-Degree survey.

Strong lensing is the observed distortion of background galaxies into arcs when their light passes through such a gravitational lens. It has been observed around many distant clusters including Abell 1689. By measuring the distortion geometry, the mass of the intervening cluster can be obtained. In the dozens of cases where this has been done, the mass-to-light ratios obtained correspond to the dynamical dark matter measurements of clusters. Lensing can lead to multiple copies of an image. By analyzing the distribution of multiple image copies, scientists have been able to deduce and map the distribution of dark matter around the MACS J0416.1-2403 galaxy cluster.

Weak gravitational lensing investigates minute distortions of galaxies, using statistical analyses from vast galaxy surveys. By examining the apparent shear deformation of the adjacent background galaxies, the mean distribution of dark matter can be characterized. The mass-to-light ratios correspond to dark matter densities predicted by other large-scale structure measurements. Dark matter does not bend light itself; mass (in this case the mass of the dark matter) bends spacetime. Light follows the curvature of spacetime, resulting in the lensing effect.

Cosmic Microwave Background

Although both dark matter and ordinary matter are matter, they do not behave in the same way. In particular, in the early universe, ordinary matter was ionized and interacted strongly with radiation via Thomson scattering. Dark matter does not interact directly with radiation, but it does affect the CMB by its gravitational potential (mainly on large scales), and by its effects on the density and velocity of ordinary matter. Ordinary and dark matter perturbations, therefore, evolve differently with time and leave different imprints on the cosmic microwave background (CMB).

The cosmic microwave background is very close to a perfect blackbody but contains very small temperature anisotropies of a few parts in 100,000. A sky map of anisotropies can be decomposed into an angular power spectrum, which is observed to contain a series of acoustic peaks at near-equal spacing but different heights. The series of peaks can be predicted for any assumed set of cosmological parameters by modern computer codes such as CMBFast and CAMB, and matching theory to data, therefore, constrains cosmological parameters. The first peak mostly shows the density of baryonic matter, while the third peak relates mostly to the density of dark matter, measuring the density of matter and the density of atoms.

The CMB anisotropy was first discovered by COBE in 1992, though this had too coarse resolution to detect the acoustic peaks. After the discovery of the first acoustic peak by the balloon-borne BOOMERanG experiment in 2000, the power spectrum was precisely observed by WMAP in 2003–2012, and even more precisely by the Planck spacecraft in 2013–2015. The results support the Lambda-CDM model.

The observed CMB angular power spectrum provides powerful evidence in support of dark matter, as its precise structure is well fitted by the Lambda-CDM model, but difficult to reproduce with any competing model such as modified Newtonian dynamics (MOND).

Structure Formation

Structure formation refers to the period after the Big Bang when density perturbations collapsed to form stars, galaxies, and clusters. Prior to structure formation, the Friedmann solutions to general relativity describe a homogeneous universe. Later, small anisotropies gradually grew and condensed the homogeneous universe into stars, galaxies and larger structures. Ordinary matter is affected by radiation, which is the dominant element of the universe at very early times. As a result, its density perturbations are washed out and unable to condense into structure. If there were only ordinary matter in the universe, there would not have been enough time for density perturbations to grow into the galaxies and clusters currently seen.

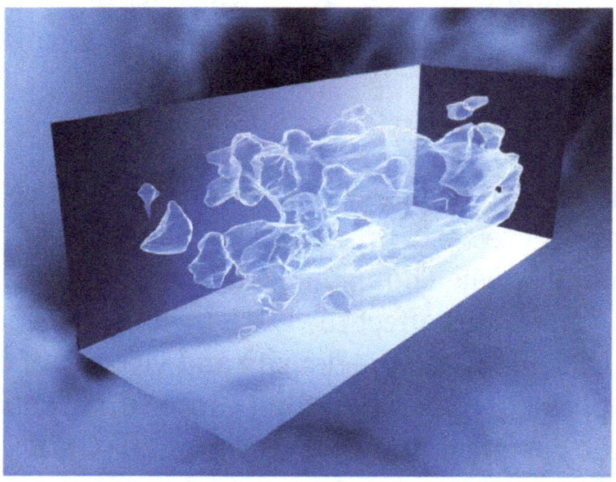

3-D map of the large-scale distribution of dark matter, reconstructed from measurements of weak gravitational lensing with the Hubble Space Telescope.

Dark matter provides a solution to this problem because it is unaffected by radiation. Therefore, its density perturbations can grow first. The resulting gravitational potential acts as an attractive potential well for ordinary matter collapsing later, speeding up the structure formation process.

Bullet Cluster

If dark matter does not exist, then the next most likely explanation must be general relativity – the prevailing theory of gravity – is incorrect and should be modified. The Bullet Cluster, the result of a recent collision of two galaxy clusters, provides a challenge for modified gravity theories because its apparent center of mass is far displaced from the baryonic center of mass. Standard dark matter models can easily explain this observation, but modified gravity has a much harder time, especially since the observational evidence is model-independent.

Type I a Supernova Distance Measurements

Type Ia supernovae can be used as standard candles to measure extragalactic distances, which can in turn be used to measure how fast the universe has expanded in the past. Data indicates the universe is expanding at an accelerating rate, the cause of which is usually ascribed to dark energy. Since observations indicate the universe is almost flat, it is expected the total energy density of everything in the universe should sum to 1 ($\Omega_{tot} \approx 1$). The measured dark energy density is $\Omega_\Lambda \approx 0.690$; the observed ordinary (baryonic) matter energy density is $\Omega_b \approx 0.0482$ and the energy density of radiation is negligible. This leaves a missing $\Omega_{dm} \approx 0.258$ which nonetheless behaves like matter – dark matter.

Sky Surveys and Baryon Acoustic Oscillations

Baryon acoustic oscillations (BAO) are fluctuations in the density of the visible baryonic matter (normal matter) of the universe on large scales. These are predicted to arise in the Lambda-CDM model due to acoustic oscillations in the photon-baryon fluid of the early universe, and can be observed in the cosmic microwave background angular power spectrum. BAOs set up a preferred length scale for baryons. As the dark matter and baryons clumped together after recombination, the effect is much weaker in the galaxy distribution in the nearby universe, but is detectable as a subtle (≈1 percent) preference for pairs of galaxies to be separated by 147 Mpc, compared to those separated by 130–160 Mpc. This feature was predicted theoretically in the 1990s and then discovered in 2005, in two large galaxy redshift surveys, the Sloan Digital Sky Survey and the 2dF Galaxy Redshift Survey. Combining the CMB observations with BAO measurements from galaxy redshift surveys provides a precise estimate of the Hubble constant and the average matter density in the Universe. The results support the Lambda-CDM model.

Redshift-space Distortions

Large galaxy redshift surveys may be used to make a three-dimensional map of the galaxy distribution. These maps are slightly distorted because distances are estimated from observed redshifts; the redshift contains a contribution from the galaxy's so-called peculiar velocity in addition to the dominant Hubble expansion term. On average, superclusters are expanding more slowly than the cosmic mean due to their gravity, while voids are expanding faster than average. In a redshift map, galaxies in front of a supercluster have excess radial velocities towards it and have redshifts slightly higher than their distance would imply, while galaxies behind the supercluster have redshifts slightly low for their distance. This effect causes superclusters to appear squashed in the radial direction, and likewise voids are stretched. Their angular positions are unaffected. This effect is not detectable for any one structure since the true shape is not known, but can be measured by averaging over many structures. It was predicted quantitatively by Nick Kaiser in 1987, and first decisively measured in 2001 by the 2dF Galaxy Redshift Survey. Results are in agreement with the Lambda-CDM model.

Lyman-alpha Forest

In astronomical spectroscopy, the Lyman-alpha forest is the sum of the absorption lines arising from the Lyman-alpha transition of neutral hydrogen in the spectra of distant galaxies and quasars. Lyman-alpha forest observations can also constrain cosmological models. These constraints agree with those obtained from WMAP data.

Composition of Dark Matter: Baryonic vs. Nonbaryonic

Dark matter can refer to any substance which interacts predominantly via gravity with visible matter (e.g., stars and planets). Hence in principle it need not be composed of a new type of fundamental particle but could, at least in part, be made up of standard baryonic matter, such as protons or neutrons. However, for the reasons outlined below, most scientists think the dark matter is dominated by a non-baryonic component, which is likely composed of a currently unknown fundamental particle (or similar exotic state).

Baryonic Matter

Baryons (protons and neutrons) make up ordinary stars and planets. However, baryonic matter also encompasses less common non-primordial black holes, neutron stars, faint old white dwarfs and brown dwarfs, collectively known as massive compact halo objects (MACHOs), which can be hard to detect.

However, multiple lines of evidence suggest the majority of dark matter is not made of baryons:

- Sufficient diffuse, baryonic gas or dust would be visible when backlit by stars.
- The theory of Big Bang nucleosynthesis predicts the observed abundance of the chemical elements. If there are more baryons, then there should also be more helium, lithium and heavier elements synthesized during the Big Bang.

Agreement with observed abundances requires that baryonic matter makes up between 4–5% of the universe's critical density. In contrast, large-scale structure and other observations indicate that the total matter density is about 30% of the critical density.

- Astronomical searches for gravitational microlensing in the Milky Way found at most only a small fraction of the dark matter may be in dark, compact, conventional objects (MACHOs, etc.); the excluded range of object masses is from half the Earth's mass up to 30 solar masses, which covers nearly all the plausible candidates.
- Detailed analysis of the small irregularities (anisotropies) in the cosmic microwave background. Observations by WMAP and Planck indicate that around five-sixths of the total matter is in a form that interacts significantly with ordinary matter or photons only through gravitational effects.

Non-baryonic Matter

Candidates for non-baryonic dark matter are hypothetical particles such as axions, sterile neutrinos, weakly interacting massive particles (WIMPs), gravitationally-interacting massive particles (GIMPs), supersymmetric particles, or primordial black holes. The three neutrino types already observed are indeed abundant, and dark, and matter, but because their individual masses – however uncertain they may be – are almost certainly too tiny, they can only supply a small fraction of dark matter, due to limits derived from large-scale structure and high-redshift galaxies.

Unlike baryonic matter, nonbaryonic matter did not contribute to the formation of the elements in the early universe (Big Bang nucleosynthesis) and so its presence is revealed only via its gravitational effects, or weak lensing. In addition, if the particles of which it is composed are supersymmetric, they can undergo annihilation interactions with themselves, possibly resulting in observable by-products such as gamma rays and neutrinos (indirect detection).

Dark Matter Aggregation and Dense Dark Matter Objects

If dark matter is composed of weakly-interacting particles, an obvious question is whether it can form objects equivalent to planets, stars, or black holes. Historically, the answer has been it cannot, because of two factors:

It lacks an efficient means to lose energy: Ordinary matter forms dense objects because it has numerous ways to lose energy. Losing energy would be essential for object formation, because a particle that gains energy during compaction or falling "inward" under gravity, and cannot lose it any other way, will heat up and increase velocity and momentum. Dark matter appears to lack means to lose energy, simply because it is not capable of interacting strongly in other ways except through gravity.

The virial theorem suggests that such a particle would not stay bound to the gradually forming object – as the object began to form and compact, the dark matter particles within it would speed up and tend to escape.

It lacks a range of interactions needed to form structures: Ordinary matter interacts in many different ways. This allows the matter to form more complex structures. For example, stars form through gravity, but the particles within them interact and can emit energy in the form of neutrinos and electromagnetic radiation through fusion when they become energetic enough. Protons and neutrons can bind via the strong interaction and then form atoms with electrons largely through electromagnetic interaction. But there is no evidence that dark matter is capable of such a wide variety of interactions, since it seems to only interact through gravity (and possibly through some means no stronger than the weak interaction, although until dark matter is better understood, this is only hopeful speculation).

In 2015–2017 the idea dense dark matter was composed of primordial black holes, made a comeback following results of gravitation wave measurements which detected the merger of intermediate mass black holes. Black holes with about 30 solar masses are not predicted to form by either stellar collapse (typically less than 15 solar masses) or by the merger of black holes in galactic centers (millions or billions of solar masses). It was proposed the intermediate mass black holes causing the detected merger formed in the hot dense early phase of the universe due to denser regions collapsing. However this was later ruled out by a survey of about a thousand supernova which detected no gravitational lensing events, although about 8 would be expected if intermediate mass primordial black holes accounted for the majority of dark matter.

The possibility atom-sized primordial black holes account for a significant fraction of dark matter was ruled out by measurements of positron and electron fluxes outside the Sun's heliosphere by the Voyager 1 spacecraft. Tiny black holes are theorized to emit Hawking radiation. However the detected fluxes were too low and did not have the expected energy spectrum suggesting tiny primordial black holes are not widespread enough to account for dark matter. Nonetheless, research and theories proposing dense dark matter accounts for dark matter continue as of 2018, including approaches to dark matter cooling, and the question remains unsettled. In 2019, the lack of microlensing effects in the observation of Andromeda suggests tiny black holes do not exist. However, there still exists a largely unconstrained mass range smaller than that can be limited by optical microlensing observations, where primordial black holes may account for all dark matter.

Classification of Dark Matter: Cold, Warm or Hot

Dark matter can be divided into *cold*, *warm*, and *hot* categories. These categories refer to velocity rather than an actual temperature, indicating how far corresponding objects moved due to random motions in the early universe, before they slowed due to cosmic expansion – this is an important distance called the *free streaming length* (FSL). Primordial density fluctuations smaller than this length get washed out as particles spread from overdense to underdense regions, while larger fluctuations are unaffected; therefore this length sets a minimum scale for later structure formation.

The categories are set with respect to the size of a protogalaxy (an object that later evolves into a dwarf galaxy): Dark matter particles are classified as cold, warm, or hot according to their FSL; much smaller (cold), similar to (warm), or much larger (hot) than a protogalaxy. Mixtures of the above are also possible: a theory of mixed dark matter was popular in the mid-1990s, but was rejected following the discovery of dark energy.

Cold dark matter leads to a bottom-up formation of structure with galaxies forming first and galaxy clusters at a latter stage, while hot dark matter would result in a top-down formation scenario with large matter aggregations forming early, later fragmenting into separate galaxies; the latter is excluded by high-redshift galaxy observations.

These categories also correspond to fluctuation spectrum effects and the interval following the Big Bang at which each type became non-relativistic. Davis wrote in 1985:

Candidate particles can be grouped into three categories on the basis of their effect on the fluctuation spectrum. If the dark matter is composed of abundant light particles which remain relativistic until shortly before recombination, then it may be termed "hot". The best candidate for hot dark matter is a neutrino. A second possibility is for the dark matter particles to interact more weakly than neutrinos, to be less abundant, and to have a mass of order 1 keV. Such particles are termed "warm dark matter", because they have lower thermal velocities than massive neutrinos there are at present few candidate particles which fit this description. Gravitinos and photinos have been suggested. Any particles which became nonrelativistic very early, and so were able to diffuse a negligible distance, are termed "cold" dark matter (CDM). There are many candidates for CDM including supersymmetric particles.

—M. Davis, G. Efstathiou, C.S. Frenk, and S.D.M. White

Another approximate dividing line is warm dark matter became non-relativistic when the universe was approximately 1 year old and 1 millionth of its present size and in the radiation-dominated era (photons and neutrinos), with a photon temperature 2.7 million Kelvins. Standard physical cosmology gives the particle horizon size as 2 $c\,t$ (speed of light multiplied by time) in the radiation-dominated era, thus 2 light-years. A region of this size would expand to 2 million light-years today (absent structure formation). The actual FSL is approximately 5 times the above length, since it continues to grow slowly as particle velocities decrease inversely with the scale factor after they become non-relativistic. In this example the FSL would correspond to 10 million light-years, or 3 mega parsecs, today, around the size containing an average large galaxy.

The 2.7 million K photon temperature gives a typical photon energy of 250 electron-volts, thereby setting a typical mass scale for warm dark matter: Particles much more massive than this, such as GeV–TeV mass WIMPs, would become non-relativistic much earlier than one year after the Big Bang and thus have FSLs much smaller than a protogalaxy, making them cold. Conversely, much lighter particles, such as neutrinos with masses of only a few eV, have FSLs much larger than a protogalaxy, thus qualifying them as hot.

Cold Dark Matter

Cold dark matter offers the simplest explanation for most cosmological observations. It is dark matter composed of constituents with an FSL much smaller than a protogalaxy. This is the focus for dark matter research, as hot dark matter does not seem capable of supporting galaxy or galaxy cluster formation, and most particle candidates slowed early.

The constituents of cold dark matter are unknown. Possibilities range from large objects like MACHOs (such as black holes and Preon stars) or RAMBOs (such as clusters of brown dwarfs), to new particles such as WIMPs and axions.

Studies of Big Bang nucleosynthesis and gravitational lensing convinced most cosmologists that MACHOs cannot make up more than a small fraction of dark matter. According to A. Peter: "The only *really plausible* dark-matter candidates are new particles."

The 1997 DAMA/NaI experiment and its successor DAMA/LIBRA in 2013, claimed to directly detect dark matter particles passing through the Earth, but many researchers remain skeptical, as negative results from similar experiments seem incompatible with the DAMA results.

Many supersymmetric models offer dark matter candidates in the form of the WIMPy Lightest Supersymmetric Particle (LSP). Separately, heavy sterile neutrinos exist in non-supersymmetric extensions to the standard model which explain the small neutrino mass through the seesaw mechanism.

Warm Dark Matter

Warm dark matter comprises particles with an FSL comparable to the size of a protogalaxy. Predictions based on warm dark matter are similar to those for cold dark matter on large scales, but with less small-scale density perturbations. This reduces the predicted abundance of dwarf galaxies and may lead to lower density of dark matter in the central parts of large galaxies. Some researchers consider this a better fit to observations. A challenge for this model is the lack of particle candidates with the required mass ≈ 300 eV to 3000 eV.

No known particles can be categorized as warm dark matter. A postulated candidate is the sterile neutrino: A heavier, slower form of neutrino that does not interact through the weak force, unlike other neutrinos. Some modified gravity theories, such as scalar–tensor–vector gravity, require "warm" dark matter to make their equations work.

Hot Dark Matter

Hot dark matter consists of particles whose FSL is much larger than the size of a protogalaxy. The neutrino qualifies as such particle. They were discovered independently, long before the hunt for dark matter: They were postulated in 1930, and detected in 1956. Neutrinos' mass is less than 10^{-6} that of an electron. Neutrinos interact with normal matter only via gravity and the weak force, making them difficult to detect (the weak force only works over a small distance, thus a neutrino triggers a weak force event only if it hits a nucleus head-on). This makes them 'weakly interacting light particles' (WILPs), as opposed to WIMPs.

The three known flavours of neutrinos are the *electron*, *muon*, and *tau*. Their masses are slightly different. Neutrinos oscillate among the flavours as they move. It is hard to determine an exact upper bound on the collective average mass of the three neutrinos (or for any of the three individually). For example, if the average neutrino mass were over 50 eV/c^2 (less than 10^{-5} of the mass of an electron), the universe would collapse. CMB data and other methods indicate that their average mass probably does not exceed 0.3 eV/c^2. Thus, observed neutrinos cannot explain dark matter.

Because galaxy-size density fluctuations get washed out by free-streaming, hot dark matter implies the first objects that can form are huge supercluster-size pancakes, which then fragment into galaxies. Deep-field observations show instead that galaxies formed first, followed by clusters and superclusters as galaxies clump together.

Detection of Dark Matter Particles

If dark matter is made up of sub-atomic particles, then millions, possibly billions, of such particles must pass through every square centimeter of the Earth each second. Many experiments aim to test this hypothesis. Although WIMPs are popular search candidates, the Axion Dark Matter Experiment (ADMX) searches for axions. Another candidate is heavy hidden sector particles which only interact with ordinary matter via gravity.

These experiments can be divided into two classes: direct detection experiments, which search for the scattering of dark matter particles off atomic nuclei within a detector; and indirect detection, which look for the products of dark matter particle annihilations or decays.

Direct Detection

Direct detection experiments aim to observe low-energy recoils (typically a few keVs) of nuclei induced by interactions with particles of dark matter, which are passing through the Earth. After such a recoil the nucleus will emit energy in the form of scintillation light or phonons, as they pass through sensitive detection apparatus. To do this effectively, it is crucial to maintain a low background, and so such experiments operate deep underground to reduce the interference from cosmic rays. Examples of underground laboratories with direct detection experiments include the Stawell mine, the Soudan mine, the SNOLAB underground laboratory at Sudbury, the Gran Sasso National Laboratory, the Canfranc Underground Laboratory, the Boulby Underground Laboratory, the Deep Underground Science and Engineering Laboratory and the China Jinping Underground Laboratory.

These experiments mostly use either cryogenic or noble liquid detector technologies. Cryogenic detectors operating at temperatures below 100 mK, detect the heat produced when a particle hits an atom in a crystal absorber such as germanium. Noble liquid detectors detect scintillation produced by a particle collision in liquid xenon or argon. Cryogenic detector experiments include: CDMS, CRESST, EDELWEISS, EURECA. Noble liquid experiments include ZEPLIN, XENON, DEAP, ArDM, WARP, DarkSide, PandaX, and LUX, the Large Underground Xenon experiment. Both of these techniques focus strongly on their ability to distinguish background particles (which predominantly scatter off electrons) from dark matter particles (that scatter off nuclei). Other experiments include SIMPLE and PICASSO.

Currently there has been no well-established claim of dark matter detection from a direct detection experiment, leading instead to strong upper limits on the mass and interaction cross section with nucleons of such dark matter particles. The DAMA/NaI and more recent DAMA/LIBRA experimental collaborations have detected an annual modulation in the rate of events in their detectors, which they claim is due to dark matter. This results from the expectation that as the Earth orbits the Sun, the velocity of the detector relative to the dark matter halo will vary by a small amount. This claim is so far unconfirmed and in contradiction with negative results from other experiments such as LUX, SuperCDMS and XENON100.

A special case of direct detection experiments covers those with directional sensitivity. This is a search strategy based on the motion of the Solar System around the Galactic Center. A low-pressure time projection chamber makes it possible to access information on recoiling tracks and constrain WIMP-nucleus kinematics. WIMPs coming from the direction in which the Sun travels (approximately towards Cygnus) may then be separated from background, which should be isotropic. Directional dark matter experiments include DMTPC, DRIFT, Newage and MIMAC.

Indirect Detection

Indirect detection experiments search for the products of the self-annihilation or decay of dark matter particles in outer space. For example, in regions of high dark matter density (e.g., the centre of our galaxy) two dark matter particles could annihilate to produce gamma rays or Standard Model particle-antiparticle pairs. Alternatively, if the dark matter particle is unstable, it could decay into Standard Model (or other) particles. These processes could be detected indirectly through an excess of gamma rays, antiprotons or positrons emanating from high density regions in our galaxy or others. A major difficulty inherent in such searches is that various astrophysical sources can mimic the signal expected from dark matter, and so multiple signals are likely required for a conclusive discovery.

Collage of six cluster collisions with dark matter maps. The clusters were observed in a study of how dark matter in clusters of galaxies behaves when the clusters collide.

A few of the dark matter particles passing through the Sun or Earth may scatter off atoms and lose energy. Thus dark matter may accumulate at the center of these bodies, increasing the chance of collision/annihilation. This could produce a distinctive signal in the form of high-energy neutrinos. Such a signal would be strong indirect proof of WIMP dark matter. High-energy neutrino telescopes such as AMANDA, IceCube and ANTARES are searching for this signal. The detection by LIGO in September 2015 of gravitational waves, opens the possibility of observing dark matter in a new way, particularly if it is in the form of primordial black holes.

Many experimental searches have been undertaken to look for such emission from dark matter annihilation or decay, examples of which follow. The Energetic Gamma Ray Experiment Telescope observed more gamma rays in 2008 than expected from the Milky Way, but scientists concluded this was most likely due to incorrect estimation of the telescope's sensitivity.

The Fermi Gamma-ray Space Telescope is searching for similar gamma rays. In April 2012, an analysis of previously available data from its Large Area Telescope instrument produced statistical evidence of a 130 GeV signal in the gamma radiation coming from the center of the Milky Way. WIMP annihilation was seen as the most probable explanation.

At higher energies, ground-based gamma-ray telescopes have set limits on the annihilation of dark matter in dwarf spheroidal galaxies and in clusters of galaxies.

The PAMELA experiment detected excess positrons. They could be from dark matter annihilation or from pulsars. No excess antiprotons were observed.

In 2013 results from the Alpha Magnetic Spectrometer on the International Space Station indicated excess high-energy cosmic rays which could be due to dark matter annihilation.

Collider Searches for Dark Matter

An alternative approach to the detection of dark matter particles in nature is to produce them in a laboratory. Experiments with the Large Hadron Collider (LHC) may be able to detect dark matter particles produced in collisions of the LHC proton beams. Because a dark matter particle should have negligible interactions with normal visible matter, it may be detected indirectly as (large amounts of) missing energy and momentum that escape the detectors, provided other (non-negligible) collision products are detected. Constraints on dark matter also exist from the LEP experiment using a similar principle, but probing the interaction of dark matter particles with electrons rather than quarks. Any discovery from collider searches must be corroborated by discoveries in the indirect or direct detection sectors to prove that the particle discovered is, in fact, dark matter.

Alternative Hypotheses

Because dark matter has not yet been conclusively identified, many other hypotheses have emerged aiming to explain the observational phenomena that dark matter was conceived to explain. The most common method is to modify general relativity. General relativity is well-tested on solar system scales, but its validity on galactic or cosmological scales has not been well proven. A suitable modification to general relativity can conceivably eliminate the need for dark matter. The best-known theories of this class are MOND and its relativistic generalization tensor-vector-scalar gravity (TeVeS), f(R) gravity, negative mass dark fluid, and entropic gravity. Alternative theories abound.

A problem with alternative hypotheses is observational evidence for dark matter comes from so many independent approaches. Explaining any individual observation is possible but explaining all of them is very difficult. Nonetheless, there have been some scattered successes for alternative hypotheses, such as a 2016 test of gravitational lensing in entropic gravity.

The prevailing opinion among most astrophysicists is while modifications to general relativity can conceivably explain part of the observational evidence, there is probably enough data to conclude there must be some form of dark matter.

Dark Matter Halo

According to modern models of physical cosmology, a dark matter halo is a basic unit of cosmological structure. It is a region that has decoupled from cosmic expansion and contains gravitationally bound matter. A single dark matter halo may contain multiple virialized clumps of dark matter bound together by gravity, known as subhalos. Modern cosmological models, such as ΛCDM, propose that dark matter halos and subhalos may contain galaxies. The dark matter halo of a galaxy envelops the galactic disc and extends well beyond the edge of the visible galaxy. Thought to consist of dark matter, halos have not been observed directly. Their existence is inferred through observations of their effects on the motions of stars and gas in galaxies and gravitational lensing. Dark matter halos play a key role in current models of galaxy formation and evolution. Theories that attempt to explain the nature of dark matter halos with varying degrees of success include Cold Dark Matter (CDM), Warm Dark Matter, and massive compact halo objects (MACHOs).

Simulated dark matter halo from a cosmological N-body simulation.

Galaxy rotation curve for the Milky Way. Vertical axis is speed of rotation about the galactic center. Horizontal axis is distance from the galactic center. The sun is marked with a yellow ball. The observed curve of speed of rotation is blue. The predicted curve based upon stellar mass and gas in the Milky Way is red. Scatter in observations roughly indicated by gray bars. The difference is due to dark matter or perhaps a modification of the law of gravity.

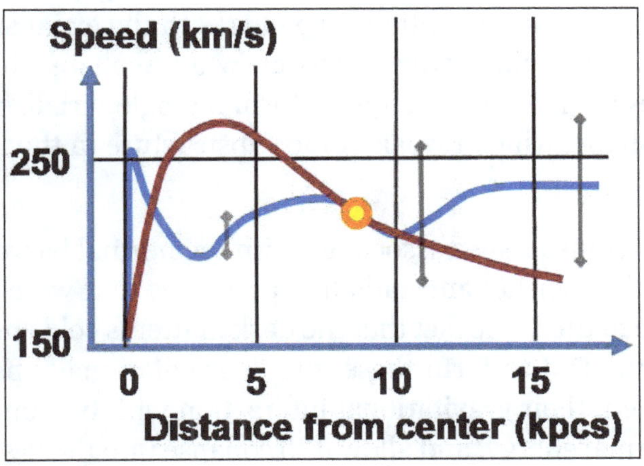

Rotation Curves as Evidence of a Dark Matter Halo

The presence of dark matter (DM) in the halo is inferred from its gravitational effect on a spiral galaxy's rotation curve. Without large amounts of mass throughout the (roughly spherical) halo, the rotational velocity of the galaxy would decrease at large distances from the galactic center, just as the orbital speeds of the outer planets decrease with distance from the Sun. However, observations of spiral galaxies, particularly radio observations of line emission from neutral atomic hydrogen (known, in astronomical parlance, as 21 cm Hydrogen line, H one, and H I line), show that the rotation curve of most spiral galaxies flattens out, meaning that rotational velocities do not decrease with distance from the galactic center. The absence of any visible matter to account for these observations implies either that unobserved ("dark") matter, first proposed by Ken Freeman in 1970, exist, or that the theory of motion under gravity (General Relativity) is incomplete. Freeman noticed that the expected decline in velocity was not present in NGC 300 nor M33, and considered an undetected mass to explain it. The DM Hypothesis has been reinforced by several studies.

Formation and Structure of Dark Matter Halos

The formation of dark matter halos is believed to have played a major role in the early formation of galaxies. During initial galactic formation, the temperature of the baryonic matter should have still been much too high for it to form gravitationally self-bound objects, thus requiring the prior formation of dark matter structure to add additional gravitational interactions. The current hypothesis for this is based on cold dark matter (CDM) and its formation into structure early in the universe.

The hypothesis for CDM structure formation begins with density perturbations in the Universe that grow linearly until they reach a critical density, after which they would stop expanding and collapse to form gravitationally bound dark matter halos. These halos would continue to grow in mass (and size), either through accretion of material from their immediate neighborhood, or by merging with other halos. Numerical simulations of CDM structure formation have been found to proceed as follows: A small volume with small perturbations initially expands with the expansion of the Universe. As time proceeds, small-scale perturbations grow and collapse to form small halos. At a later stage, these small halos merge to form a single virialized dark matter halo with an ellipsoidal shape, which reveals some substructure in the form of dark matter sub-halos.

The use of CDM overcomes issues associated with the normal baryonic matter because it removes most of the thermal and radiative pressures that were preventing the collapse of the baryonic matter. The fact that the dark matter is cold compared to the baryonic matter allows the DM to form these initial, gravitationally bound clumps. Once these subhalos formed, their gravitational interaction with baryonic matter is enough to overcome the thermal energy, and allow it to collapse into the first stars and galaxies.

Simulations of this early galaxy formation matches the structure observed by galactic surveys as well as observation of the Cosmic Microwave Background.

Density Profiles

A commonly used model for galactic dark matter halos is the pseudo-isothermal halo:

$$\rho(r) = \rho_o \left[1 + \left(\frac{r}{r_c}\right)^2\right]^{-1}$$

where ρ_o denotes the finite central density and r_c the core radius. This provides a good fit to most rotation curve data. However, it cannot be a complete description, as the enclosed mass fails to converge to a finite value as the radius tends to infinity. The isothermal model is, at best, an approximation. Many effects may cause deviations from the profile predicted by this simple model. For example, (i) collapse may never reach an equilibrium state in the outer region of a dark matter halo, (ii) non-radial motion may be important, and (iii) mergers associated with the (hierarchical) formation of a halo may render the spherical-collapse model invalid.

Numerical simulations of structure formation in an expanding universe lead to the empirical NFW (Navarro-Frenk-White) profile:

$$\rho(r) = \frac{\rho_{crit}\delta_c}{\left(\frac{r}{r_s}\right)\left(1 + \frac{r}{r_s}\right)^2}$$

where r_s is a scale radius, δ_c is a characteristic (dimensionless) density, and $\rho_{crit} = 3H^2 / 8\pi G$ is the critical density for closure. The NFW profile is called 'universal' because it works for a large variety of halo masses, spanning four orders of magnitude, from individual galaxies to the halos of galaxy clusters. This profile has a finite gravitational potential even though the integrated mass still diverges logarithmically. It has become conventional to refer to the mass of a halo at a fiducial point that encloses an overdensity 200 times greater than the critical density of the universe, though mathematically the profile extends beyond this notational point. It was later deduced that the density profile depends on the environment, with the NFW appropriate only for isolated halos. NFW halos generally provide a worse description of galaxy data than does the pseudo-isothermal profile, leading to the cuspy halo problem.

Higher resolution computer simulations are better described by the Einasto profile:

$$\rho(r) = \rho_e Exp\left[-d_n\left(\left(\frac{r}{r_c}\right)^{\frac{1}{n}} - 1\right)\right]$$

where r is the spatial (i.e., not projected) radius. The term d_n is a function of n such that ρ_e is the density at the radius r_e that defines a volume containing half of the total mass. While the addition of a third parameter provides a slightly improved description of the results from numerical simulations, it is not observationally distinguishable from the 2 parameter NFW halo, and does nothing to alleviate the cuspy halo problem.

Shape

The collapse of overdensities in the cosmic density field is generally aspherical. So, there is no reason to expect the resulting halos to be spherical. Even the earliest simulations of structure formation in a CDM universe emphasized that the halos are substantially flattened. Subsequent work has shown that halo equidensity surfaces can be described by ellipsoids characterized by the lengths of their axes.

Because of uncertainties in both the data and the model predictions, it is still unclear whether the halo shapes inferred from observations are consistent with the predictions of ΛCDM cosmology.

Halo Substructure

Up until the end of the 1990s, numerical simulations of halo formation revealed little substructure. With increasing computing power and better algorithms, it became possible to use greater numbers of particles and obtain better resolution. Substantial amounts of substructure are now expected. When a small halo merges with a significantly larger halo it becomes a subhalo orbiting within the potential well of its host. As it orbits, it is subjected to strong tidal forces from the host, which cause it to lose mass. In addition the orbit itself evolves as the subhalo is subjected to dynamical friction which causes it to lose energy and angular momentum to the dark matter particles of its host. Whether a subhalo survives as a self-bound entity depends on its mass, density profile, and its orbit.

Angular Momentum

As originally pointed out by Hoyle and first demonstrated using numerical simulations by Efstathiou & Jones, asymmetric collapse in an expanding universe produces objects with significant angular momentum.

Numerical simulations have shown that the spin parameter distribution for halos formed by dissipation-less hierarchical clustering is well fit by a log-normal distribution, the median and width of which depend only weakly on halo mass, redshift, and cosmology:

$$\rho(\lambda)d\lambda = \frac{1}{\sqrt{2\pi}\sigma_{ln\lambda}} \exp\left[-\frac{\ln\left(\frac{\lambda}{\bar{\lambda}}\right)^2}{2\sigma^2_{\ln\lambda}}\right]\frac{d\lambda}{\lambda}$$

with $\bar{\lambda} \approx 0.035$ and $\sigma_{ln\lambda} \approx 0.5$. At all halo masses, there is a marked tendency for halos with higher spin to be in denser regions and thus to be more strongly clustered.

Milky Way Dark Matter Halo

The visible disk of the Milky Way Galaxy is embedded in a much larger, roughly spherical halo of dark matter. The dark matter density drops off with distance from the galactic center. It is now believed that about 95% of the Galaxy is composed of dark matter, a type of matter that does not seem to interact with the rest of the Galaxy's matter and energy in any way except through gravity. The luminous matter makes up approximately 9×10^{10} solar masses. The dark matter halo is likely to include around 6×10^{11} to 3×10^{12} solar masses of dark matter.

Shape of the Universe

The shape of the universe is the local and global geometry of the universe. The local features of the geometry of the universe are primarily described by its curvature, whereas the topology of the universe describes general global properties of its shape as of a continuous object. The shape of the universe is related to general relativity, which describes how spacetime is curved and bent by mass and energy.

Cosmologists distinguish between the observable universe and the global universe. The observable universe consists of the part of the universe that can, in principle, be observed by light reaching Earth within the age of the universe. It encompasses a region of space that currently forms a ball centered at Earth of estimated radius 46.5 billion light-years (4.40×10^{26} m). This does not mean the universe is 46.5 billion years old; instead the universe is measured to be 13.8 billion years old, but space itself has also expanded, causing the size of the observable universe to be larger than the distance traversible by light over the duration of its current age. Assuming an isotropic nature, the observable universe is similar for all contemporary vantage points.

The global shape of the universe can be described with three attributes:

- Finite or infinite.
- Flat (no curvature), open (negative curvature), or closed (positive curvature).
- Connectivity, how the universe is put together, i.e., simply connected space or multiply connected.

There are certain logical connections among these properties. For example, a universe with positive curvature is necessarily finite. Although it is usually assumed in the literature that a flat or negatively curved universe is infinite, this need not be the case if the topology is not the trivial one: For example, a three-torus is flat but finite.

The exact shape is still a matter of debate in physical cosmology, but experimental data from various independent sources (WMAP, BOOMERanG, and Planck for example) confirm that the universe is flat with only a 0.4% margin of error. Theorists have been trying to construct a formal mathematical model of the shape of the universe. In formal terms, this is a 3-manifold model corresponding to the spatial section (in comoving coordinates) of the 4-dimensional spacetime of the universe. The model most theorists currently use is the Friedmann–Lemaître–Robertson–Walker (FLRW) model. Arguments have been put forward that the observational data best fit with the conclusion that the shape of the global universe is infinite and flat, but the data are also consistent with other possible shapes, such as the so-called Poincaré dodecahedral space and the Sokolov–Starobinskii space (quotient of the upper half-space model of hyperbolic space by 2-dimensional lattice).

Shape of the Observable Universe

As stated in the introduction, there are two aspects to consider:

- Its local geometry, which predominantly concerns the curvature of the universe, particularly the observable universe.
- Its global geometry, which concerns the topology of the universe as a whole.

The observable universe can be thought of as a sphere that extends outwards from any observation point for 46.5 billion light years, going farther back in time and more redshifted the more distant away one looks. Ideally, one can continue to look back all the way to the Big Bang; in practice, however, the farthest away one can look using light and other electromagnetic radiation is the cosmic microwave background (CMB), as anything past that was opaque. Experimental investigations show that the observable universe is very close to isotropic and homogeneous.

If the observable universe encompasses the entire universe, we may be able to determine the structure of the entire universe by observation. However, if the observable universe is smaller than the entire universe, our observations will be limited to only a part of the whole, and we may not be able to determine its global geometry through measurement. From experiments, it is possible to construct different mathematical models of the global geometry of the entire universe, all of which are consistent with current observational data; thus it is currently unknown whether the observable universe is identical to the global universe, or is instead many orders of magnitude smaller. The universe may be small in some dimensions and not in others (analogous to the way a cuboid is longer in the dimension of length than it is in the dimensions of width and depth). To test whether a given mathematical model describes the universe accurately, scientists look for the model's novel implications—what are some phenomena in the universe that we have not yet observed, but that must exist if the model is correct—and they devise experiments to test whether those phenomena occur or not. For example, if the universe is a small closed loop, one would expect to see multiple images of an object in the sky, although not necessarily images of the same age.

Cosmologists normally work with a given space-like slice of spacetime called the comoving coordinates, the existence of a preferred set of which is possible and widely accepted in present-day physical cosmology. The section of spacetime that can be observed is the backward light cone (all points within the cosmic light horizon, given time to reach a given observer), while the related term Hubble volume can be used to describe either the past light cone or comoving space up to the surface of last scattering. To speak of "the shape of the universe (at a point in time)" is ontologically naive from the point of view of special relativity alone: due to the relativity of simultaneity we cannot speak of different points in space as being "at the same point in time" nor, therefore, of "the shape of the universe at a point in time". However, the comoving coordinates (if well-defined) provide a strict sense to those by using the time since the Big Bang (measured in the reference of CMB) as a distinguished universal time.

Curvature of the Universe

The curvature is a quantity describing how the geometry of a space differs locally from the one of the flat space. The curvature of any locally isotropic space (and hence of a locally isotropic universe) falls into one of the three following cases:

- Zero curvature (flat); a drawn triangle's angles add up to 180° and the Pythagorean theorem holds; such 3-dimensional space is locally modeled by Euclidean space E^3.
- Positive curvature; a drawn triangle's angles add up to more than 180°; such 3-dimensional space is locally modeled by a region of a 3-sphere S^3.
- Negative curvature; a drawn triangle's angles add up to less than 180°; such 3-dimensional space is locally modeled by a region of a hyperbolic space H^3.

Curved geometries are in the domain of Non-Euclidean geometry. An example of a positively curved space would be the surface of a sphere such as the Earth. A triangle drawn from the equator to a pole will have at least two angles equal 90°, which makes the sum of the 3 angles greater than 180°. An example of a negatively curved surface would be the shape of a saddle or mountain pass. A triangle drawn on a saddle surface will have the sum of the angles adding up to less than 180°.

The local geometry of the universe is determined by whether the density parameter Ω is greater than, less than, or equal to 1. From top to bottom: a spherical universe with

$\Omega > 1$, a hyperbolic universe with $\Omega < 1$, and a flat universe with $\Omega = 1$. These depictions of two-dimensional surfaces are merely easily visualizable analogs to the 3-dimensional structure of (local) space.

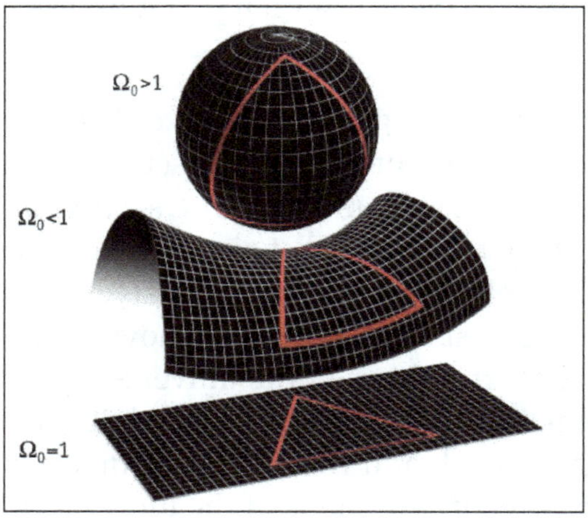

General relativity explains that mass and energy bend the curvature of spacetime and is used to determine what curvature the universe has by using a value called the density parameter, represented with Omega (Ω). The density parameter is the average density of the universe divided by the critical energy density, that is, the mass energy needed for a universe to be flat. Put another way,

- If $\Omega = 1$, the universe is flat.
- If $\Omega > 1$, there is positive curvature.
- if $\Omega < 1$ there is negative curvature.

One can experimentally calculate this Ω to determine the curvature two ways. One is to count up all the mass-energy in the universe and take its average density then divide that average by the critical energy density. Data from Wilkinson Microwave Anisotropy Probe (WMAP) as well as the Planck spacecraft give values for the three constituents of all the mass-energy in the universe – normal mass (baryonic matter and dark matter), relativistic particles (photons and neutrinos), and dark energy or the cosmological constant:

$$\Omega_{mass} \approx 0.315 \pm 0.018.$$

$$\Omega_{relativistic} \approx 9.24 \times 10^{-5}.$$

$$\Omega_{\Lambda} \approx 0.6817 \pm 0.0018.$$

$$\Omega_{total} = \Omega_{mass} + \Omega_{relativistic} + \Omega_{\Lambda} = 1.00 \pm 0.02.$$

The actual value for critical density value is measured as $\rho_{critical} = 9.47 \times 10^{-27}$ kg m^{-3}. From these values, within experimental error, the universe seems to be flat.

Another way to measure Ω is to do so geometrically by measuring an angle across the observable universe. We can do this by using the CMB and measuring the power spectrum and temperature anisotropy. For an intuition, one can imagine finding a gas cloud that is not in thermal equilibrium due to being so large that light speed cannot propagate the thermal information. Knowing this propagation speed, we then know the size of the gas cloud as well as the distance to the gas cloud, we then have two sides of a triangle and can then determine the angles. Using a method similar to this, the BOOMERanG experiment has determined that the sum of the angles to 180° within experimental error, corresponding to an $\Omega_{total} \approx 1.00 \pm 0.12$.

These and other astronomical measurements constrain the spatial curvature to be very close to zero, although they do not constrain its sign. This means that although the local geometries of spacetime are generated by the theory of relativity based on spacetime intervals, we can approximate 3-space by the familiar Euclidean geometry.

The Friedmann–Lemaître–Robertson–Walker (FLRW) model using Friedmann equations is commonly used to model the universe. The FLRW model provides a curvature of the universe based on the mathematics of fluid dynamics, that is, modeling the matter within the universe as a perfect fluid. Although stars and structures of mass can be introduced into an "almost FLRW" model, a strictly FLRW model is used to approximate the local geometry of the observable universe. Another way of saying this is that if all forms of dark energy are ignored, then the curvature of the universe can be determined by measuring the average density of matter within it, assuming that all matter is evenly distributed (rather than the distortions caused by 'dense' objects such as galaxies). This assumption is justified by the observations that, while the universe is "weakly" inhomogeneous and anisotropic, it is on average homogeneous and isotropic.

Global Universe Structure

Global structure covers the geometry and the topology of the whole universe—both the observable universe and beyond. While the local geometry does not determine the global geometry completely, it does limit the possibilities, particularly a geometry of a constant curvature. The universe is often taken to be a geodesic manifold, free of topological defects; relaxing either of these complicates the analysis considerably. A global geometry is a local geometry plus a topology. It follows that a topology alone does not give a global geometry: for instance, Euclidean 3-space and hyperbolic 3-space have the same topology but different global geometries.

As stated in the introduction, investigations within the study of the global structure of the universe include:

- Whether the universe is infinite or finite in extent.
- Whether the geometry of the global universe is flat, positively curved, or negatively curved.

- Whether the topology is simply connected like a sphere or multiply connected, like a torus.

Infinite or Finite

One of the presently unanswered questions about the universe is whether it is infinite or finite in extent. For intuition, it can be understood that a finite universe has a finite volume that, for example, could be in theory filled up with a finite amount of material, while an infinite universe is unbounded and no numerical volume could possibly fill it. Mathematically, the question of whether the universe is infinite or finite is referred to as boundedness. An infinite universe (unbounded metric space) means that there are points arbitrarily far apart: for any distance d, there are points that are of a distance at least d apart. A finite universe is a bounded metric space, where there is some distance d such that all points are within distance d of each other. The smallest such d is called the diameter of the universe, in which case the universe has a well-defined "volume" or "scale."

With or without Boundary

Assuming a finite universe, the universe can either have an edge or no edge. Many finite mathematical spaces, e.g., a disc, have an edge or boundary. Spaces that have an edge are difficult to treat, both conceptually and mathematically. Namely, it is very difficult to state what would happen at the edge of such a universe. For this reason, spaces that have an edge are typically excluded from consideration.

However, there exist many finite spaces, such as the 3-sphere and 3-torus, which have no edges. Mathematically, these spaces are referred to as being compact without boundary. The term compact basically means that it is finite in extent ("bounded") and complete. The term "without boundary" means that the space has no edges. Moreover, so that calculus can be applied, the universe is typically assumed to be a differentiable manifold. A mathematical object that possesses all these properties, compact without boundary and differentiable, is termed a closed manifold. The 3-sphere and 3-torus are both closed manifolds.

Curvature

The curvature of the universe places constraints on the topology. If the spatial geometry is spherical, i.e., possess positive curvature, the topology is compact. For a flat (zero curvature) or a hyperbolic (negative curvature) spatial geometry, the topology can be either compact or infinite. Many textbooks erroneously state that a flat universe implies an infinite universe; however, the correct statement is that a flat universe that is also simply connected implies an infinite universe. For example, Euclidean space is flat, simply connected, and infinite, but the torus is flat, multiply connected, finite, and compact.

In general, local to global theorems in Riemannian geometry relate the local geometry to the global geometry. If the local geometry has constant curvature, the global geometry is very constrained, as described in Thurston geometries.

The latest research shows that even the most powerful future experiments (like the SKA) will not be able to distinguish between flat, open and closed universe if the true value of cosmological curvature parameter is smaller than 10^{-4}. If the true value of the cosmological curvature parameter is larger than 10^{-3} we will be able to distinguish between these three models even now.

Results of the Planck mission released in 2015 show the cosmological curvature parameter, Ω_K, to be 0.000±0.005, consistent with a flat universe.

Universe with Zero Curvature

In a universe with zero curvature, the local geometry is flat. The most obvious global structure is that of Euclidean space, which is infinite in extent. Flat universes that are finite in extent include the torus and Klein bottle. Moreover, in three dimensions, there are 10 finite closed flat 3-manifolds, of which 6 are orientable and 4 are non-orientable. These are the Bieberbach manifolds. The most familiar is the aforementioned 3-torus universe.

In the absence of dark energy, a flat universe expands forever but at a continually decelerating rate, with expansion asymptotically approaching zero. With dark energy, the expansion rate of the universe initially slows down, due to the effect of gravity, but eventually increases. The ultimate fate of the universe is the same as that of an open universe.

A flat universe can have zero total energy.

Universe with Positive Curvature

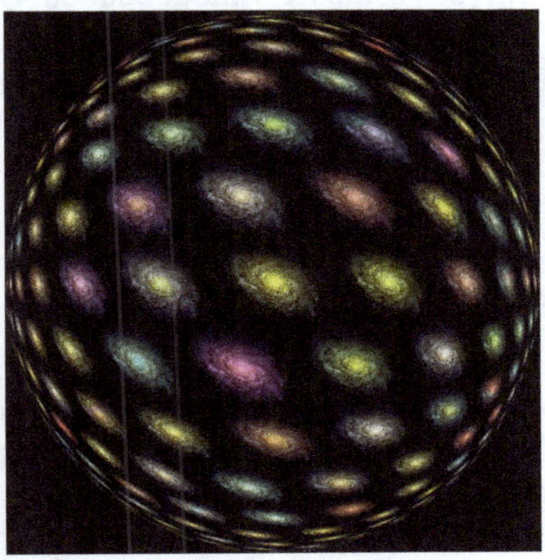

Universe in an expanding sphere. The galaxies farthest away are moving fastest and hence experience length contraction and so become smaller to an observer in the centre.

A positively curved universe is described by elliptic geometry, and can be thought of as a three-dimensional hypersphere, or some other spherical 3-manifold (such as the Poincaré dodecahedral space), all of which are quotients of the 3-sphere.

Poincaré dodecahedral space is a positively curved space, colloquially described as "soccerball-shaped", as it is the quotient of the 3-sphere by the binary icosahedral group, which is very close to icosahedral symmetry, the symmetry of a soccer ball. This was proposed by Jean-Pierre Luminet and colleagues in and an optimal orientation on the sky for the model was estimated in 2008.

Universe with Negative Curvature

A hyperbolic universe, one of a negative spatial curvature, is described by hyperbolic geometry, and can be thought of locally as a three-dimensional analog of an infinitely extended saddle shape. There are a great variety of hyperbolic 3-manifolds, and their classification is not completely understood. Those of finite volume can be understood via the Mostow rigidity theorem. For hyperbolic local geometry, many of the possible three-dimensional spaces are informally called "horn topologies", so called because of the shape of the pseudosphere, a canonical model of hyperbolic geometry. An example is the Picard horn, a negatively curved space, colloquially described as "funnel-shaped".

Curvature: Open or Closed

When cosmologists speak of the universe as being "open" or "closed", they most commonly are referring to whether the curvature is negative or positive. These meanings of open and closed are different from the mathematical meaning of open and closed used for sets in topological spaces and for the mathematical meaning of open and closed manifolds, which gives rise to ambiguity and confusion. In mathematics, there are definitions for a closed manifold (i.e., compact without boundary) and open manifold (i.e., one that is not compact and without boundary). A "closed universe" is necessarily a closed manifold. An "open universe" can be either a closed or open manifold. For example, in the Friedmann–Lemaître–Robertson–Walker (FLRW) model the universe is considered to be without boundaries, in which case "compact universe" could describe a universe that is a closed manifold.

Milne Model ("Spherical" Expanding)

If one applies Minkowski space-based special relativity to expansion of the universe, without resorting to the concept of a curved spacetime, then one obtains the Milne model. Any spatial section of the universe of a constant age (the proper time elapsed from the Big Bang) will have a negative curvature; this is merely a pseudo-Euclidean geometric fact analogous to one that concentric spheres in the flat Euclidean space are nevertheless curved. Spatial geometry of this model is an unbounded hyperbolic space. The entire universe is contained within a light cone, namely the future cone of the Big Bang. For any given moment t > 0 of coordinate time (assuming the Big Bang has t = 0), the entire universe is bounded by a sphere of radius exactly ct. The apparent paradox of an infinite universe contained within a sphere is explained with length contraction: the galaxies farther away, which are travelling away from the observer the fastest, will appear thinner.

This model is essentially a degenerate FLRW for $\Omega = 0$. It is incompatible with observations that definitely rule out such a large negative spatial curvature. However, as a background in which gravitational fields (or gravitons) can operate, due to diffeomorphism invariance, the space on the macroscopic scale, is equivalent to any other (open) solution of Einstein's field equations.

Reionization

In the field of Big Bang theory, and cosmology, reionization is the process that caused the matter in the universe to reionize after the lapse of the "dark ages".

Reionization is the second of two major phase transitions of gas in the universe. While the majority of baryonic matter in the universe is in the form of hydrogen and helium, reionization usually refers strictly to the reionization of hydrogen, the element.

It's believed that the primordial helium also experienced the same phase of reionization changes, but at different points in the history of the universe. This is usually referred to as helium reionization.

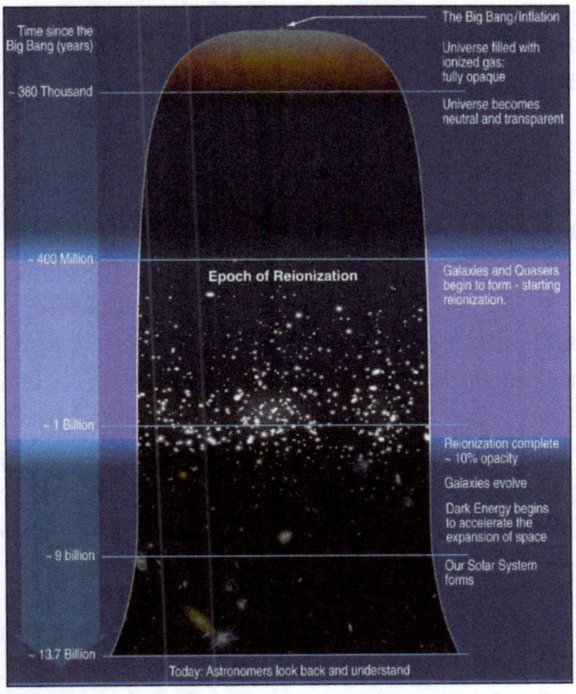

Schematic timeline of the universe, depicting reionization's place in cosmic history.

The first phase change of hydrogen in the universe was recombination, which occurred at a redshift z = 1089 (379,000 years after the Big Bang), due to the cooling of the universe to the point where the rate of recombination of electrons and protons to form neutral hydrogen was higher than the reionization rate. The universe was opaque before the recombination, due to the scattering of photons (of all wavelengths) off free electrons (and, to a significantly lesser extent, free protons), but it became increasingly transparent as more electrons and protons combined to form neutral hydrogen atoms. While the electrons of neutral hydrogen can absorb photons of some wavelengths by rising to an excited state, a universe full of neutral hydrogen will be relatively opaque only at those absorbed wavelengths, but transparent throughout most of the spectrum. The Dark Ages of the universe start at that point, because there were no light sources other than the gradually redshifting cosmic background radiation.

The second phase change occurred once objects started to condense in the early universe that were energetic enough to re-ionize neutral hydrogen. As these objects formed and radiated energy, the universe reverted from being neutral, to once again being an ionized plasma. This occurred between 150 million and one billion years after the Big Bang (at a redshift $6 < z < 20$). At that time, however, matter had been diffused by the expansion of the universe, and the scattering interactions of photons and electrons were much less frequent than before electron-proton recombination. Thus, the universe was full of low density ionized hydrogen and remained transparent, as is the case today.

Detection Methods

Looking back so far in the history of the universe presents some observational challenges. There are, however, a few observational methods for studying reionization.

Quasars and the Gunn-peterson Trough

One means of studying reionization uses the spectra of distant quasars. Quasars release an extraordinary amount of energy, in fact they are among the brightest objects in the universe. As a result, some quasars are detectable from as far back as the epoch of reionization. Quasars also happen to have relatively uniform spectral features, regardless of their position in the sky or distance from the Earth. Thus it can be inferred that any major differences between quasar spectra will be caused by the interaction of their emission with atoms along the line of sight. For wavelengths of light at the energies of one of the Lyman transitions of hydrogen, the scattering cross-section is large, meaning that even for low levels of neutral hydrogen in the intergalactic medium (IGM), absorption at those wavelengths is highly likely.

For nearby objects in the universe, spectral absorption lines are very sharp, as only photons with energies just sufficient to cause an atomic transition can cause that transition. However, the distances between quasars and the telescopes which detect them are large, which means that the expansion of the universe causes light to undergo noticeable redshifting. This means that as light from the quasar travels through the IGM and is redshifted, wavelengths which had been below the Lyman Alpha limit are stretched, and will in effect begin to fill in the Lyman absorption band. This means that instead of showing sharp spectral absorption lines, a quasar's light which has traveled through a large, spread out region of neutral hydrogen will show a Gunn-Peterson trough.

The redshifting for a particular quasar provides temporal (time) information about reionization. Since an object's redshift corresponds to the time at which it emitted the light, it is possible to determine when reionization ended. Quasars below a certain redshift (closer in space and time) do not show the Gunn-Peterson trough (though they may show the Lyman-alpha forest), while quasars emitting light prior to reionization will feature a Gunn-Peterson trough. In 2001, four quasars were detected (by the Sloan Digital Sky Survey) with redshifts ranging from $z = 5.82$ to $z = 6.28$. While the quasars above $z = 6$ showed a Gunn-Peterson trough, indicating that the IGM was still at least partly neutral, the ones below did not, meaning the hydrogen was ionized. As reionization is expected to occur over relatively short timescales, the results suggest that the universe was approaching the end of reionization at $z = 6$. This, in turn, suggests that the universe must still have been almost entirely neutral at $z > 10$.

CMB Anisotropy and Polarization

The anisotropy of the cosmic microwave background on different angular scales can also be used to study reionization. Photons undergo scattering when there are free electrons present, in a process known as Thomson scattering. However, as the universe expands, the density of free electrons will decrease, and scattering will occur less frequently. In the period during and after reionization, but before significant expansion had occurred to sufficiently lower the electron density, the light that composes the CMB will experience observable Thomson scattering. This scattering will leave its mark on the CMB anisotropy map, introducing secondary anisotropies (anisotropies introduced after recombination). The overall effect is to erase anisotropies that occur on smaller scales. While anisotropies on small scales are erased, polarization anisotropies are actually introduced because of reionization. By looking at the CMB anisotropies observed, and comparing with what they would look like had reionization not taken place, the electron column density at the time of reionization can be determined. With this, the age of the universe when reionization occurred can then be calculated.

The Wilkinson Microwave Anisotropy Probe allowed that comparison to be made. The initial observations, released in 2003, suggested that reionization took place from $11 < z < 30$. This redshift range was in clear disagreement with the results from studying quasar spectra. However, the three year WMAP data returned a different result, with reionization beginning at $z = 11$ and the universe ionized by $z = 7$. This is in much better agreement with the quasar data.

Results in 2018 from Planck mission, yield an instantaneous reionization redshift of z = 7.68 ± 0.79.

The parameter usually quoted here is τ, the "optical depth to recombination," or alternatively, z_{re}, the redshift of reionization, assuming it was an instantaneous event. While this is unlikely to be physical, since reionization was very likely not instantaneous, z_{re} provides an estimate of the mean redshift of reionization.

21-cm Line

Even with the quasar data roughly in agreement with the CMB anisotropy data, there are still a number of questions, especially concerning the energy sources of reionization and the effects on, and role of, structure formation during reionization. The 21-cm line in hydrogen is potentially a means of studying this period, as well as the "dark ages" that preceded reionization. The 21-cm line occurs in neutral hydrogen, due to differences in energy between the spin triplet and spin singlet states of the electron and proton. This transition is forbidden, meaning it occurs extremely rarely. The transition is also highly temperature dependent, meaning that as objects form in the "dark ages" and emit Lyman-alpha photons that are absorbed and re-emitted by surrounding neutral hydrogen, it will produce a 21-cm line signal in that hydrogen through Wouthuysen-Field coupling. By studying 21-cm line emission, it will be possible to learn more about the early structures that formed. Observations from the Experiment to Detect the Global Epoch of Reionization Signature (EDGES) points to a signal from this era, although follow-up observations will be needed to confirm it. Several other projects hope to make headway in this area in the near future, such as the Precision Array for Probing the Epoch of Reionization (PAPER), Low Frequency Array (LOFAR), Murchison Widefield Array (MWA), Giant Metrewave Radio Telescope (GMRT), the Dark Ages Radio Explorer (DARE) mission, and the Large-Aperture Experiment to Detect the Dark Ages (LEDA).

Energy Sources

While observations have come in which narrow the window during which the epoch of reionization could have taken place, it is still uncertain which objects provided the photons that reionized the IGM. To ionize neutral hydrogen, an energy larger than 13.6 eV is required, which corresponds to photons with a wavelength of 91.2 nm or shorter. This is in the ultraviolet part of the electromagnetic spectrum, which means that the primary candidates are all sources which produce a significant amount of energy in the ultraviolet and above. How numerous the source is must also be considered, as well as the longevity, as protons and electrons will recombine if energy is not continuously provided to keep them apart. Altogether, the critical parameter for any source considered can be summarized as its "emission rate of hydrogen-ionizing photons per unit cosmological volume." With these constraints, it is expected that quasars and first generation stars and galaxies were the main sources of energy.

Astronomers hope to use observations to answer the question of how the Universe was reionised.

Dwarf Galaxies

Dwarf galaxies are currently the primary source of ionizing photons during the epoch of reionization. For most scenarios, this would require the log-slope of the UV galaxy luminosity function, often denoted α, to be steeper than it is today, approaching α = -2.

In 2014, two separate sources have identified two Green Pea galaxies (GPs) to be likely Lyman Continuum (LyC)-emitting candidates. This suggests that these two GPs are low-redshift analogs of high-redshift Lyman-alpha and LyC emitters, only two others of which are known: Haro 11 and Tololo-1247-232. Finding local LyC emitters is crucial to the theories about the early universe and the epoch of reionization. These two GPs have SDSS DR9 reference numbers: 1237661070336852109 (GP_J1219) and 1237664668421849521.

A new study shows that dwarf galaxies contributed nearly 30% of the ultraviolet light during the process of reionization. The dwarfs had such a big impact because a larger fraction of ionizing photons are able to escape dwarf galaxies (clocking in at 50%) as opposed to larger galaxies (clocking in at a mere 5%). Quoting J.H. Wise from an interview with Sky and Telescope: "The smallest galaxies first dominate at early times; however, they basically kill themselves off by blowing out their gas through their own supernovae and heating their environment. Afterwards, larger galaxies (but still much smaller than the Milky Way by about 100 times in mass) take over the job of reionizing the universe."

Quasars

Quasars, a class of active galactic nuclei (AGN), were considered a good candidate source because they are highly efficient at converting mass to energy, and emit a great deal of light above the threshold for ionizing hydrogen. It is unknown, however, how many quasars existed prior to reionization. Only the brightest of quasars present during reionization can be detected, which means there is no direct information about dimmer quasars that existed. However, by looking at the more easily observed quasars in the nearby universe, and assuming that the luminosity function (number of quasars as a function of luminosity) during reionization will be approximately the same as it is today, it is possible to make estimates of the quasar populations at earlier times. Such studies have found that quasars do not exist in high enough numbers to reionize the IGM alone, saying that "only if the ionizing background is dominated by low-luminosity AGNs can the quasar luminosity function provide enough ionizing photons."

Population III Stars

Population III stars were the earliest stars, which had no elements more massive than hydrogen or helium. During Big Bang nucleosynthesis, the only elements that formed aside from hydrogen and helium were trace amounts of lithium. Yet quasar spectra have revealed the presence of heavy elements in the intergalactic medium at an early era. Supernova explosions produce such heavy elements, so hot, large, Population III stars which will form supernovae are a possible mechanism for reionization. While they have not been directly observed, they are consistent according to models using numerical simulation and current observations. A gravitationally lensed galaxy also provides indirect evidence of Population III stars. Even without direct observations of Population III stars, they are a compelling source. They are more efficient and effective ionizers than Population II stars, as they emit more ionizing photons, and are capable of reionizing hydrogen on their own in some reionization models with reasonable initial mass functions. As a consequence, Population III stars are currently considered the most likely energy source to initiate the reionization of the universe, though other sources are likely to have taken over and driven reionization to completion.

Simulated image of the first stars, 400 Myr after the Big Bang.

In June 2015, astronomers reported evidence for Population III stars in the Cosmos Redshift 7 galaxy at $z = 6.60$. Such stars are likely to have existed in the very early universe (i.e., at high redshift), and may have started the production of chemical elements heavier than hydrogen that are needed for the later formation of planets and life as we know it.

Electroweak Epoch

In physical cosmology the electroweak epoch was the period in the evolution of the early universe when the temperature of the universe was high enough to merge electromagnetism and the weak interaction into a single electroweak interaction (> 100 GeV). The electroweak epoch began approximately 10^{-36} seconds after the Big Bang, when the strong force separated from the electroweak interaction. This phase transition triggered a period of exponential expansion known as cosmic inflation. At approximately 10^{-32} seconds after the Big Bang the potential energy of the inflaton field that had driven the inflation of the universe during the inflationary epoch was released, filling the universe with a dense, hot quark-gluon plasma. Particle interactions in this phase were energetic enough to create large numbers of exotic particles, including W and Z bosons and Higgs bosons. As the universe expanded and cooled, interactions became less energetic and when the universe was about 10^{-12} seconds old, W and Z bosons ceased to be created. The remaining W and Z bosons decayed quickly, and the weak interaction became a short-range force in the following quark epoch.

After the inflationary period, the physics of the electroweak epoch is less speculative and much better understood than the physics of previous periods of the early universe. The existence of W and Z bosons has been demonstrated, and other predictions of electroweak theory have been experimentally verified.

References

- Physical-cosmology, terms: sciencedaily.com, Retrieved 18 June, 2020
- Musser, George (2001). "Copernican Counterrevolution". Scientific American. 284 (3): 24. Bibcode:2001SciAm.284c..24M. doi:10.1038/scientificamerican0301-24a
- William C. Keel (2007). The Road to Galaxy Formation (2nd ed.). Springer-Praxis. ISBN 978-3-540-72534-3
- Cosmic Neutrinos Detected, Confirming The Big Bang's Last Great Prediction - Forbes coverage of original paper: Follin, Brent; Knox, Lloyd; Millea, Marius; Pan, Zhen (2015). "First Detection of the Acoustic Oscillation Phase Shift Expected from the Cosmic Neutrino Background". Physical Review Letters. 115 (9): 091301. arXiv:1503.07863. Bibcode:2015PhRvL.115i1301F. doi:10.1103/PhysRevLett.115.091301. PMID 26371637
- Electroweak-Epoch, Astronomy-Cosmology: scientificlib.com, Retrieved 19 May, 2020
- Bertone, G.; Hooper, D.; Silk, J. (2005). "Particle dark matter: Evidence, candidates and constraints". Physics Reports. 405 (5–6): 279–390. arXiv:hep-ph/0404175. Bibcode:2005PhR...405..279B. doi:10.1016/j.physrep.2004.08.031
- Wechsler, Risa; Tinker, Jeremy (September 2018). "The Connection between Galaxies and their Dark Matter Halos". Annual Review of Astronomy and Astrophysics. 56: 435–487. arXiv:1804.03097. doi:10.1146/annurev-astro-081817-051756
- R.H. Becker; et al. (2001). "Evidence For Reionization at z ~ 6: Detection of a Gunn-Peterson Trough In A z=6.28 Quasar". Astronomical Journal. 122 (6): 2850–2857. arXiv:astro-ph/0108097. Bibcode:2001AJ....122.2850B. doi:10.1086/324231

4

The Big Bang

The Big Bang is a cosmological model that describes the evolution and expansion of universe from a very high density and high-temperature state. Lambda-CDM model, cosmological horizon, timeline of the universe, etc. are a few of its aspects. This chapter sheds light on the subject of the Big Bang for a thorough understanding of the subject.

The Big Bang Theory is the leading explanation about how the universe began. At its simplest, it says the universe as we know it started with a small singularity, then inflated over the next 13.8 billion years to the cosmos that we know today.

Because current instruments don't allow astronomers to peer back at the universe's birth, much of what we understand about the Big Bang Theory comes from mathematical formulas and models. Astronomers can, however, see the "echo" of the expansion through a phenomenon known as the cosmic microwave background.

While the majority of the astronomical community accepts the theory, there are some theorists who have alternative explanations besides the Big Bang — such as eternal inflation or an oscillating universe.

The phrase "Big Bang Theory" has been popular among astrophysicists for decades, but it hit the mainstream in 2007 when a comedy show with the same name premiered on CBS. The show follows the home and academic life of several researchers (including an astrophysicist).

In the first second after the universe began, the surrounding temperature was about 10 billion degrees Fahrenheit (5.5 billion Celsius), according to NASA. The cosmos contained a vast array of fundamental particles such as neutrons, electrons and protons. These decayed or combined as the universe got cooler.

This early soup would have been impossible to look at, because light could not carry inside of it. "The free electrons would have caused light (photons) to scatter the way sunlight scatters from the water droplets in clouds," NASA stated. Over time, however, the free electrons met up with nuclei and created neutral atoms. This allowed light to shine through about 380,000 years after the Big Bang.

This early light — sometimes called the "afterglow" of the Big Bang — is more properly known as the cosmic microwave background (CMB). It was first predicted by Ralph Alpher and other scientists in 1948, but was found only by accident almost 20 years later.

Arno Penzias and Robert Wilson, both of Bell Telephone Laboratories in Murray Hill, New Jersey, were building a radio receiver in 1965 and picking up higher-than-expected temperatures, according to NASA. At first, they thought the anomaly was due to pigeons and their dung, but even after cleaning up the mess and killing pigeons that tried to roost inside the antenna, the anomaly persisted.

Simultaneously, a Princeton University team (led by Robert Dicke) was trying to find evidence of the CMB, and realized that Penzias and Wilson had stumbled upon it. The teams each published papers in the Astrophysical Journal in 1965.

Determining the Age of the Universe

The cosmic microwave background has been observed on many missions. One of the most famous space-faring missions was NASA's Cosmic Background Explorer (COBE) satellite, which mapped the sky in the 1990s.

Several other missions have followed in COBE's footsteps, such as the BOOMERanG experiment (Balloon Observations of Millimetric Extragalactic Radiation and Geophysics), NASA's Wilkinson Microwave Anisotropy Probe (WMAP) and the European Space Agency's Planck satellite.

Planck's observations, first released in 2013, mapped the background in unprecedented detail and revealed that the universe was older than previously thought: 13.82 billion years old, rather than 13.7 billion years old.

The maps give rise to new mysteries, however, such as why the Southern Hemisphere appears slightly redder (warmer) than the Northern Hemisphere. The Big Bang Theory says that the CMB would be mostly the same, no matter where you look.

Examining the CMB also gives astronomers clues as to the composition of the universe. Researchers think most of the cosmos is made up of matter and energy that cannot be "sensed" with conventional instruments, leading to the names dark matter and dark energy. Only 5 percent of the universe is made up of matter such as planets, stars and galaxies.

Gravitational Waves Controversy

While astronomers could see the universe's beginnings, they've also been seeking out proof of its rapid inflation. Theory says that in the first second after the universe was born, our cosmos ballooned faster than the speed of light. That, by the way, does not violate Albert Einstein's speed limit since he said that light is the maximum anything can travel within the universe. That did not apply to the inflation of the universe itself.

In 2014, astronomers said they had found evidence in the CMB concerning "B-modes," a sort of polarization generated as the universe got bigger and created gravitational waves. The team spotted evidence of this using an Antarctic telescope called "Background Imaging of Cosmic Extragalactic Polarization", or BICEP2.

"We're very confident that the signal that we're seeing is real, and it's on the sky," lead researcher John Kovac, of the Harvard-Smithsonian Center for Astrophysics, told Space.com in March 2014. But by June, the same team said that their findings could have been altered by galactic dust getting in the way of their field of view.

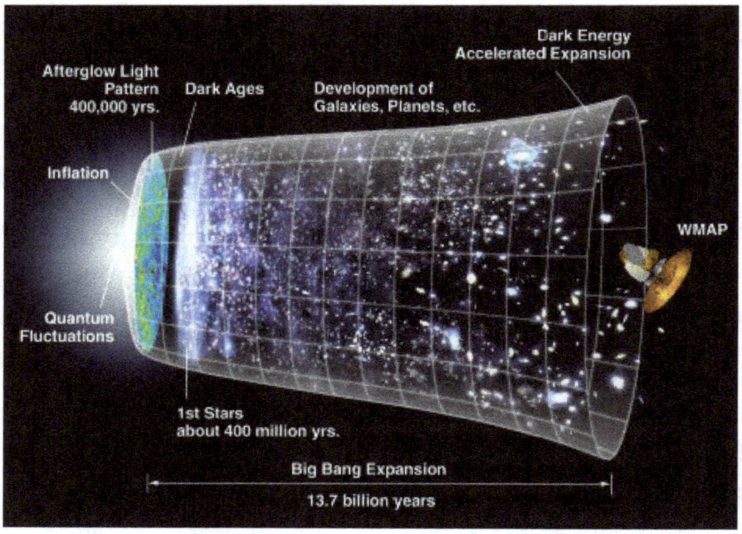

This graphic shows a timeline of the universe based on the Big Bang theory and inflation models.

Separately, gravitational waves have been confirmed when talking about the movements and collisions of black holes that are a few tens of masses larger than our sun. These waves have been detected multiple times by the Laser Interferometer Gravitational-Wave Observatory (LIGO) since 2016. As LIGO becomes more sensitive, it is anticipated that discovering black hole-related gravitational waves will be a fairly frequent event.

Faster Inflation, Multiverses and Charting the Start

The universe is not only expanding, but getting faster as it inflates. This means that with time, nobody will be able to spot other galaxies from Earth, or any other vantage point within our galaxy.

"So, if you wait long enough, eventually, a distant galaxy will reach the speed of light. What that means is that even light won't be able to bridge the gap that's being opened between that galaxy and us. There's no way for extraterrestrials on that galaxy to communicate with us, to send any signals that will reach us, once their galaxy is moving faster than light relative to us."

Some physicists also suggest that the universe we experience is just one of many. In the "multiverse" model, different universes would coexist with each other like bubbles lying side by side. The theory suggests that in that first big push of inflation, different parts of space-time grew at different rates. This could have carved off different sections – different universes – with potentially different laws of physics.

Lambda-CDM Model

The ΛCDM (Lambda cold dark matter) or Lambda-CDM model is a parametrization of the Big Bang cosmological model in which the universe contains three major components: first, a cosmological constant denoted by Lambda (Greek **Λ**) and associated with dark energy; second, the postulated cold dark matter (abbreviated CDM); and third, ordinary matter. It is frequently referred to as the standard model of Big Bang cosmology because it is the simplest model that provides a reasonably good account of the following properties of the cosmos:

- The existence and structure of the cosmic microwave background.
- The large-scale structure in the distribution of galaxies.
- The observed abundances of hydrogen (including deuterium), helium, and lithium.
- The accelerating expansion of the universe observed in the light from distant galaxies and supernovae.

The model assumes that general relativity is the correct theory of gravity on cosmological scales. It emerged in the late 1990s as a concordance cosmology, after a period of time when disparate observed properties of the universe appeared mutually inconsistent, and there was no consensus on the makeup of the energy density of the universe.

The ΛCDM model can be extended by adding cosmological inflation, quintessence and other elements that are current areas of speculation and research in cosmology.

Some alternative models challenge the assumptions of the ΛCDM model. Examples of these are modified Newtonian dynamics, entropic gravity, modified gravity, theories of large-scale variations in the matter density of the universe, bimetric gravity, scale invariance of empty space, and decaying dark matter (DDM).

Most modern cosmological models are based on the cosmological principle, which states that our observational location in the universe is not unusual or special; on a large-enough scale, the universe looks the same in all directions (isotropy) and from every location (homogeneity).

The model includes an expansion of metric space that is well documented both as the red shift of prominent spectral absorption or emission lines in the light from distant galaxies and as the time dilation in the light decay of supernova luminosity curves. Both effects are attributed to a Doppler shift in electromagnetic radiation as it travels across expanding space. Although this expansion increases the distance between objects that are not under shared gravitational influence, it does not increase the size of the objects (e.g. galaxies) in space. It also allows for distant galaxies to recede from each other at speeds greater than the speed of light; local expansion is less than the speed of light, but expansion summed across great distances can collectively exceed the speed of light.

The letter Λ (lambda) represents the cosmological constant, which is currently associated with a vacuum energy or dark energy in empty space that is used to explain the contemporary accelerating expansion of space against the attractive effects of gravity. A cosmological constant has negative pressure, $p = -\rho c^2$ which contributes to the stress-energy tensor that, according to the general theory of relativity, causes accelerating expansion. The fraction of the total energy density of our (flat or almost flat) universe that is dark energy, Ω_Λ is estimated to be 0.669 ± 0.038 based on the 2018 Dark Energy Survey results using Type Ia Supernovae or 0.6847 ± 0.0073 based on the 2018 release of *Planck* satellite data, or more than 68.3% (2018 estimate) of the mass-energy density of the universe.

Dark matter is postulated in order to account for gravitational effects observed in very large-scale structures (the "flat" rotation curves of galaxies; the gravitational lensing of light by galaxy clusters; and enhanced clustering of galaxies) that cannot be accounted for by the quantity of observed matter.

Cold dark matter as currently hypothesized is:

- Non-baryonic: It consists of matter other than protons and neutrons (and electrons, by convention, although electrons are not baryons).
- Cold: Its velocity is far less than the speed of light at the epoch of radiation-matter equality (thus neutrinos are excluded, being non-baryonic but not cold).
- Dissipationless: It cannot cool by radiating photons.
- Collisionless: The dark matter particles interact with each other and other particles only through gravity and possibly the weak force.

Dark matter constitutes about 26.8% (2013 estimate) of the mass-energy density of the universe. The remaining 4.8% (2013 estimate) comprises all ordinary matter observed as atoms, chemical elements, gas and plasma, the stuff of which visible planets, stars and galaxies are made. The great majority of ordinary matter in the universe is unseen, since visible stars and gas inside galaxies and clusters account for less than 10% of the ordinary matter contribution to the mass-energy density of the universe.

Also, the energy density includes a very small fraction (~ 0.01%) in cosmic microwave background radiation, and not more than 0.5% in relic neutrinos. Although very small today, these were much more important in the distant past, dominating the matter at redshift > 3200.

The model includes a single originating event, the "Big Bang", which was not an explosion but the abrupt appearance of expanding space-time containing radiation at temperatures of around 10^{15} K. This was immediately (within 10^{-29} seconds) followed by an exponential expansion of space by a scale multiplier of 10^{27} or more, known as cosmic inflation. The early universe remained hot (above 10,000 K) for several hundred thousand years, a state that is detectable as a residual cosmic microwave background, or CMB, a very low energy radiation emanating from all parts of the sky. The "Big Bang" scenario, with cosmic inflation and standard particle physics, is the only current cosmological model consistent with the observed continuing expansion of space, the observed distribution of lighter elements in the universe (hydrogen, helium, and lithium), and the spatial texture of minute irregularities (anisotropies) in the CMB radiation. Cosmic inflation also addresses the "horizon problem" in the CMB; indeed, it seems likely that the universe is larger than the observable particle horizon.

The model uses the Friedmann–Lemaître–Robertson–Walker metric, the Friedmann equations and the cosmological equations of state to describe the observable universe from right after the inflationary epoch to present and future.

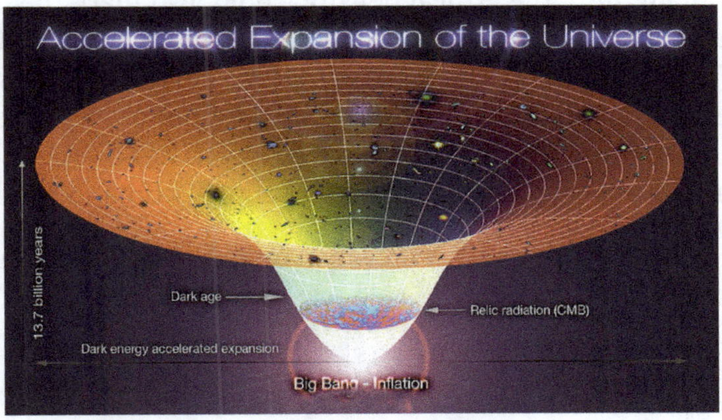

Lambda-CDM, accelerated expansion of the universe. The time-line in this schematic diagram extends from the Big Bang/inflation era 13.7 Byr ago to the present cosmological time.

The expansion of the universe is parameterized by a dimensionless scale factor $a = a(t)$ (with time t counted from the birth of the universe), defined relative to the present day, so $a_0 = a(t_0) = 1$; the usual convention in cosmology is that subscript 0 denotes present-day values, so t_0 is the current age of the universe. The scale factor is related to the observed redshift z of the light emitted at time t_{em} by:

$$a(t_{em}) = \frac{1}{1+z}.$$

The expansion rate is described by the time-dependent Hubble parameter, $H(t)$, defined as:

$$H(t) \equiv \frac{\dot{a}}{a},$$

where $\dot{a}$ is the time-derivative of the scale factor. The first Friedmann equation gives the expansion rate in terms of the matter+radiation density ρ, the curvature k, and the cosmological constant Λ:

$$H^2 = \left(\frac{\dot{a}}{a}\right)^2 = \frac{8\pi G}{3}\rho - \frac{kc^2}{a^2} + \frac{\Lambda c^2}{3},$$

where as usual c is the speed of light and G is the gravitational constant. A critical density ρ_{crit} is the present-day density, which gives zero curvature k, assuming the cosmological constant is zero, regardless of its actual value. Substituting these conditions to the Friedmann equation gives:

$$\rho_{\text{crit}} = \frac{3H_0^2}{8\pi G} = 1.878\,47(23) \times 10^{-26}\, h^2\, \text{kg}\,\text{m}^{-3},$$

where $h \equiv H_0 / (100\,\text{km}\,\text{s}^{-1}\,\text{Mpc}^{-1})$ is the reduced Hubble constant. If the cosmological constant were actually zero, the critical density would also mark the dividing line between eventual recollapse of the universe to a Big Crunch, or unlimited expansion. For the Lambda-CDM model with a positive cosmological constant (as observed), the universe is predicted to expand forever regardless of whether the total density is slightly above or below the critical density; though other outcomes are possible in extended models where the dark energy is not constant but actually time-dependent.

It is standard to define the present-day density parameter Ω_x for various species as the dimensionless ratio:

$$\Omega_x \equiv \frac{\rho_x(t = t_0)}{\rho_{\text{crit}}} = \frac{8\pi G \rho_x(t = t_0)}{3H_0^2}$$

where the subscript x is one of b for baryons, c for cold dark matter, *rad* for radiation (photons plus relativistic neutrinos), and DE or Λ for dark energy.

Since the densities of various species scale as different powers of a, e.g. a^{-3} for matter etc., the Friedmann equation can be conveniently rewritten in terms of the various density parameters as:

$$H(a) \equiv \frac{\dot{a}}{a} = H_0 \sqrt{(\Omega_c + \Omega_b)a^{-3} + \Omega_{\text{rad}} a^{-4} + \Omega_k a^{-2} + \Omega_{DE} a^{-3(1+w)}}$$

where w is the equation of state parameter of dark energy, and assuming negligible neutrino mass (significant neutrino mass requires a more complex equation). The various Ω parameters add up to 1 by construction. In the general case this is integrated by computer to give the expansion history $a(t)$ and also observable distance-redshift relations for any chosen values of the cosmological parameters, which can then be compared with observations such as supernovae and baryon acoustic oscillations.

In the minimal 6-parameter Lambda-CDM model, it is assumed that curvature Ω_k is zero and $w = -1$, so this simplifies to:

$$H(a) = H_0\sqrt{\Omega_m a^{-3} + \Omega_{\text{rad}} a^{-4} + \Omega_\Lambda}$$

Observations show that the radiation density is very small today, $\Omega_{\text{rad}} \sim 10^{-4}$; if this term is neglected the above has an analytic solution:

$$a(t) = (\Omega_m / \Omega_\Lambda)^{1/3} \sinh^{2/3}(t / t_\Lambda)$$

where $t_\Lambda \equiv 2/(3H_0\sqrt{\Omega_\Lambda})$; this is fairly accurate for $a > 0.01$ or $t > 10$ million years. Solving for $a(t) = 1$ gives the present age of the universe t_0 in terms of the other parameters.

It follows that the transition from decelerating to accelerating expansion (the second derivative $\ddot{a}$ crossing zero) occurred when:

$$a = (\Omega_m / 2\Omega_\Lambda)^{1/3}$$

which evaluates to $a \sim 0.6$ or $z \sim 0.66$ for the best-fit parameters estimated from the *Planck* spacecraft.

The discovery of the cosmic microwave background (CMB) in 1964 confirmed a key prediction of the Big Bang cosmology. From that point on, it was generally accepted that the universe started in a hot, dense state and has been expanding over time. The rate of expansion depends on the types of matter and energy present in the universe, and in particular, whether the total density is above or below the so-called critical density.

During the 1970s, most attention focused on pure-baryonic models, but there were serious challenges explaining the formation of galaxies, given the small anisotropies in the CMB (upper limits at that time). In the early 1980s, it was realized that this could be resolved if cold dark matter dominated over the baryons, and the theory of cosmic inflation motivated models with critical density.

During the 1980s, most research focused on cold dark matter with critical density in matter, around 95% CDM and 5% baryons: these showed success at forming galaxies and clusters of galaxies, but problems remained; notably, the model required a Hubble constant lower than preferred by observations, and observations around 1988–1990 showed more large-scale galaxy clustering than predicted.

These difficulties sharpened with the discovery of CMB anisotropy by the Cosmic Background Explorer in 1992, and several modified CDM models, including ΛCDM and mixed cold and hot dark matter, came under active consideration through the mid-1990s. The ΛCDM model then became the leading model following the observations of accelerating expansion in 1998, and was quickly supported by other observations: in 2000, the BOOMERanG microwave background experiment measured the total (matter–energy) density to be close to 100% of critical, whereas in 2001 the 2dFGRS galaxy redshift survey measured the matter density to be near 25%; the large difference between these values supports a positive Λ or dark energy. Much more precise spacecraft measurements of the microwave background from WMAP in 2003–2010 and *Planck* in 2013–2015 have continued to support the model and pin down the parameter values, most of which are now constrained below 1 percent uncertainty.

There is currently active research into many aspects of the ΛCDM model, both to refine the parameters and possibly detect deviations. In addition, ΛCDM has no explicit physical theory for the origin or physical nature of dark matter or dark energy; the nearly scale-invariant spectrum of the CMB perturbations, and their image across the celestial sphere, are believed to result from very small thermal and acoustic irregularities at the point of recombination.

A large majority of astronomers and astrophysicists support the ΛCDM model or close relatives of it, but Milgrom, McGaugh, and Kroupa are leading critics, attacking the dark matter portions of the theory from the perspective of galaxy formation models and supporting the alternative modified Newtonian dynamics (MOND) theory, which requires a modification of the Einstein field equations and the Friedmann equations as seen in proposals such as modified gravity theory (MOG theory) or tensor–vector–scalar gravity theory (TeVeS theory). Other proposals by theoretical astrophysicists of cosmological alternatives to Einstein's general relativity that attempt to account for dark energy or dark matter include f(R) gravity, scalar–tensor theories such as galileon theories, brane cosmologies, the DGP model, and massive gravity and its extensions such as bimetric gravity.

Successes

In addition to explaining pre-2000 observations, the model has made a number of successful predictions: notably the existence of the baryon acoustic oscillation feature, discovered in 2005 in the predicted location; and the statistics of weak gravitational lensing, first observed in 2000 by several teams. The polarization of the CMB, discovered in 2002 by DASI, is now a dramatic success: in the 2015 *Planck* data release, there are seven observed peaks in the temperature (TT) power spectrum, six peaks in the temperature-polarization (TE) cross spectrum, and five peaks in the polarization (EE) spectrum. The six free parameters can be well constrained by the TT spectrum alone, and then the TE and EE spectra can be predicted theoretically to few-percent precision with no further adjustments allowed: comparison of theory and observations shows an excellent match.

Challenges

Extensive searches for dark matter particles have so far shown no well-agreed detection; the dark energy may be almost impossible to detect in a laboratory, and its value is unnaturally small compared to naive theoretical predictions.

Comparison of the model with observations is very successful on large scales (larger than galaxies, up to the observable horizon), but may have some problems on sub-galaxy scales, possibly predicting too many dwarf galaxies and too much dark matter in the innermost regions of galaxies. This problem is called the "small scale crisis". These small scales are harder to resolve in computer simulations, so it is not yet clear whether the problem is the simulations, non-standard properties of dark matter, or a more radical error in the model.

It has been argued that the ΛCDM model is built upon a foundation of conventionalist stratagems, rendering it unfalsifiable in the sense defined by Karl Popper.

Parameters

Table: Planck Collaboration Cosmological parameters.

	Description	Symbol	Value
Indep-en-dent pa-ra-meters	Physical baryon density parameter	$\Omega_b h^2$	0.02230±0.00014
	Physical dark matter density parameter	$\Omega_c h^2$	0.1188±0.0010
	Age of the universe	t_0	13.799±0.021 × 10^9 years
	Scalar spectral index	n_s	0.9667±0.0040
	Curvature fluctuation amplitude, k_0 = 0.002 Mpc^{-1}	Δ_R^2	$2.441^{+0.088}_{-0.092} \times 10^{-9}$
	Reionization optical depth	τ	0.066±0.012
Fixed pa-ra-meters	Total density parameter	Ω_{tot}	1
	Equation of state of dark energy	w	−1
	Tensor/scalar ratio	r	0
	Running of spectral index	$dn_s/d\ln k$	0
	Sum of three neutrino masses	$\sum m_\nu$	0.06 eV/c^2
	Effective number of relativistic degrees of freedom	N_{eff}	3.046

Calcu-lated values	Hubble constant	H_0	67.74±0.46 km s^{-1} Mpc^{-1}
	Baryon density parameter	Ω_b	0.0486±0.0010
	Dark matter density parameter	Ω_c	0.2589±0.0057
	Matter density parameter	Ω_m	0.3089±0.0062
	Dark energy density parameter	Ω_Λ	0.6911±0.0062
	Critical density	ρ_{crit}	(8.62±0.12)×10^{-27} kg/m^3
	The present root-mean-square matter fluctuation averaged over a sphere of radius $8h^{-1}$ Mpc	σ_8	0.8159±0.0086
	Redshift at decoupling	z_*	1089.90±0.23
	Age at decoupling	t_*	377700±3200 years
	Redshift of reionization (with uniform prior)	z_{re}	$8.5^{+1.0}_{-1.1}$

The simple ΛCDM model is based on six parameters: physical baryon density parameter; physical dark matter density parameter; the age of the universe; scalar spectral index; curvature fluctuation amplitude; and reionization optical depth. In accordance with Occam's razor, six is the smallest number of parameters needed to give an acceptable fit to current observations; other possible parameters are fixed at "natural" values, e.g. total density parameter = 1.00, dark energy equation of state = −1.

The values of these six parameters are mostly not predicted by current theory (though, ideally, they may be related by a future "Theory of Everything"), except that most versions of cosmic inflation predict the scalar spectral index should be slightly smaller than 1, consistent with the estimated value 0.96. The parameter values, and uncertainties, are estimated using large computer searches to locate the region of parameter space providing an acceptable match to cosmological observations. From these six parameters, the other model values, such as the Hubble constant and the dark energy density, can be readily calculated.

Commonly, the set of observations fitted includes the cosmic microwave background anisotropy, the brightness/redshift relation for supernovae, and large-scale galaxy clustering including the baryon acoustic oscillation feature. Other observations, such as the Hubble constant, the abundance of galaxy clusters, weak gravitational lensing and globular cluster ages, are generally consistent with these, providing a check of the model, but are less precisely measured at present.

Parameter values listed below are from the *Planck* Collaboration Cosmological parameters 68% confidence limits for the base ΛCDM model from *Planck* CMB power spectra, in combination with lensing reconstruction and external data (BAO + JLA + H_0).

- The "physical baryon density parameter" $\Omega b\ h^2$ is the "baryon density parameter" Ωb multiplied by the square of the reduced Hubble constant $h = H_0$ / (100 km s−1 Mpc−1). Likewise for the difference between "physical dark matter density parameter" and "dark matter density parameter".

- A density ρx = Ωxρ$_{crit}$ is expressed in terms of the critical density ρ$_{crit}$, which is the total density of matter/energy needed for the universe to be spatially flat. Measurements indicate that the actual total density ρtot is very close if not equal to this value.
- This is the minimal value allowed by solar and terrestrial neutrino oscillation experiments.
- From the Standard Model of particle physics.
- Calculated from $\Omega_{bh}{}^2$ and $h = H_0 / (100 \text{ km s}^{-1} \text{ Mpc}^{-1})$.
- Calculated from $\Omega_{ch}{}^2$ and $h = H_0 / (100 \text{ km s}^{-1} \text{ Mpc}^{-1})$.
- Calculated from $h = H_0 / (100 \text{ km s}^{-1} \text{ Mpc}^{-1})$ per $\rho_{crit} = 1.87847\times10^{-26}\, h^2 \text{ kg m}^{-3}$.

Missing Baryon Problem

Massimo Persic and Paolo Salucci first estimated the baryonic density today present in ellipticals, spirals, groups and clusters of galaxies. They performed an integration of the baryonic mass-to-light ratio over luminosity (in the following M_b / L), weighted with the luminosity function $\phi(L)$ over the previously mentioned classes of astrophysical objects:

$$\rho_b = \sum \int L\phi(L) \frac{M_b}{L} dL.$$

The result was:

$$\Omega_b = \Omega_* + \Omega_{gas} = 2.2\times10^{-3} + 1.5\times10^{-3} h^{-1.3} \simeq 0.003$$

where $h \simeq 0.72$.

Note that this value is much lower than the prediction of standard cosmic nucleosynthesis $\Omega_b \simeq 0.0486$, so that stars and gas in galaxies and in galaxy groups and clusters account for less than 10% of the primordially synthesized baryons. This issue is known as the problem of the "missing baryons".

Extended Models

Table: Extended Model Parameters.

Description	Symbol	Value
Total density parameter	Ω_{tot}	1.0023+0.0056 −0.0054

Equation of state of dark energy	w	−0.980±0.053
Tensor-to-scalar ratio	r	< 0.11, k_0 = 0.002 Mpc^{-1} (2σ)
Running of the spectral index	$dn_s / d\ln k$	−0.022±0.020, k_0 = 0.002 Mpc^{-1}
Sum of three neutrino masses	$\sum m_\nu$	< 0.58 eV/c^2 (2σ)
Physical neutrino density parameter	$\Omega_\nu h^2$	< 0.0062

Extended models allow one or more of the "fixed" parameters above to vary, in addition to the basic six; so these models join smoothly to the basic six-parameter model in the limit that the additional parameter(s) approach the default values. For example, possible extensions of the simplest ΛCDM model allow for spatial curvature (Ω_{tot} may be different from 1); or quintessence rather than a cosmological constant where the equation of state of dark energy is allowed to differ from −1. Cosmic inflation predicts tensor fluctuations (gravitational waves). Their amplitude is parameterized by the tensor-to-scalar ratio (denoted r), which is determined by the unknown energy scale of inflation. Other modifications allow hot dark matter in the form of neutrinos more massive than the minimal value, or a running spectral index; the latter is generally not favoured by simple cosmic inflation models.

Allowing additional variable parameter(s) will generally *increase* the uncertainties in the standard six parameters quoted above, and may also shift the central values slightly. The Table below shows results for each of the possible "6+1" scenarios with one additional variable parameter; this indicates that, as of 2015, there is no convincing evidence that any additional parameter is different from its default value.

Some researchers have suggested that there is a running spectral index, but no statistically significant study has revealed one. Theoretical expectations suggest that the tensor-to-scalar ratio r should be between 0 and 0.3, and the latest results are now within those limits.

Age of the Universe

In physical cosmology, the age of the universe is the time elapsed since the Big Bang. The current measurement of the age of the universe is around 13.8*10^9 years– 13.799±0.021 billion years within the Lambda-CDM concordance model. The uncertainty has been narrowed down to 20 million years, based on a number of studies which all gave extremely similar figures for the age. These include studies of the microwave background radiation, and measurements by the *Planck* spacecraft, the Wilkinson Microwave Anisotropy Probe and other space probes. Measurements of the cosmic background radiation give the cooling time of the universe since the Big Bang, and measurements of the expansion rate of the universe can be used to calculate its approximate age by extrapolating backwards in time.

The Lambda-CDM concordance model describes the evolution of the universe from a very uniform, hot, dense primordial state to its present state over a span of about 13.8 billion years of cosmological time. This model is well understood theoretically and strongly supported by recent high-precision astronomical observations such as WMAP. In contrast, theories of the origin of the primordial state remain very speculative. If one extrapolates the Lambda-CDM model backward from the earliest well-understood state, it quickly (within a small fraction of a second) reaches a singularity. This is known as the "initial singularity" or the "Big Bang singularity". This singularity is not understood as having a physical significance in the usual sense, but it is convenient to quote times measured "since the Big Bang" even though they do not correspond to a physically measurable time. For example, "10^{-6} seconds after the Big Bang" is a well-defined era in the universe's evolution. If one referred to the same era as "13.8 billion years minus 10^{-6} seconds ago", the precision of the meaning would be lost because the minuscule latter time interval is eclipsed by uncertainty in the former.

Though the universe might in theory have a longer history, the International Astronomical Union presently use "age of the universe" to mean the duration of the Lambda-CDM expansion, or equivalently the elapsed time since the Big Bang in the current observable universe.

Observational Limits

Since the universe must be at least as old as the oldest things in it, there are a number of observations which put a lower limit on the age of the universe; these include the temperature of the coolest white dwarfs, which gradually cool as they age, and the dimmest turnoff point of main sequence stars in clusters (lower-mass stars spend a greater amount of time on the main sequence, so the lowest-mass stars that have evolved off of the main sequence set a minimum age).

Cosmological Parameters

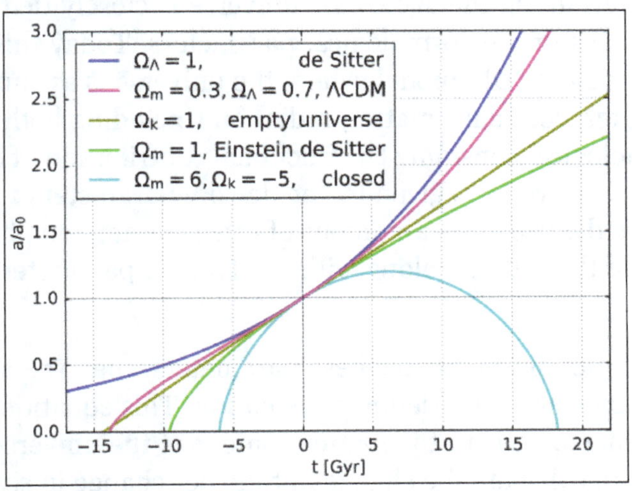

The age of the universe can be determined by measuring the Hubble constant today and extrapolating back in time with the observed value of density parameters (Ω). Before the discovery of dark energy, it was believed that the universe was matter-dominated (Einstein–de Sitter universe, green curve). Note that the de Sitter universe has infinite age, while the closed universe has the least age.

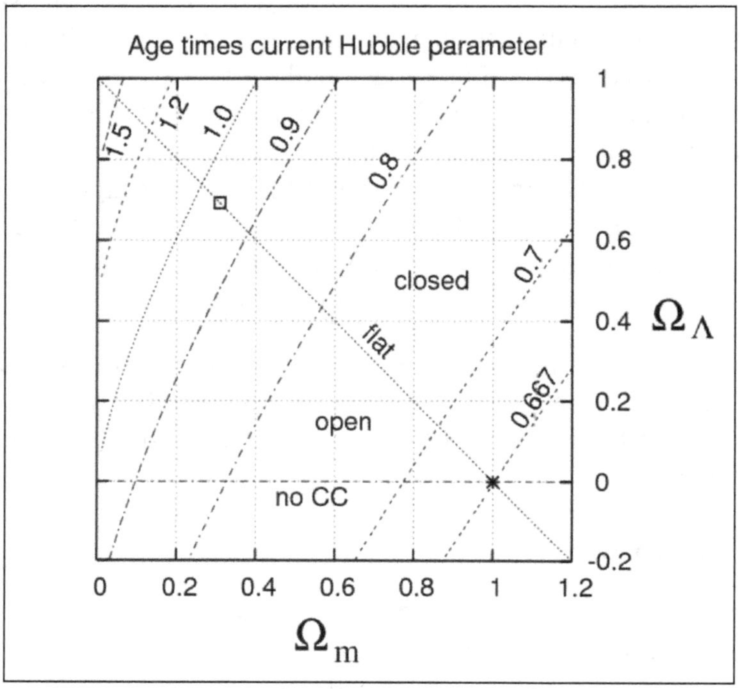

The value of the age correction factor, *F*, is shown as a function of two cosmological parameters: the current fractional matter density Ω_m and cosmological constant density Ω_Λ. The best-fit values of these parameters are shown by the box in the upper left; the matter-dominated universe is shown by the star in the lower right.

The problem of determining the age of the universe is closely tied to the problem of determining the values of the cosmological parameters. Today this is largely carried out in the context of the ΛCDM model, where the universe is assumed to contain normal (baryonic) matter, cold dark matter, radiation (including both photons and neutrinos), and a cosmological constant. The fractional contribution of each to the current energy density of the universe is given by the density parameters Ω_m, Ω_r, and Ω_Λ. The full ΛCDM model is described by a number of other parameters, but for the purpose of computing its age these three, along with the Hubble parameter H_0, are the most important.

If one has accurate measurements of these parameters, then the age of the universe can be determined by using the Friedmann equation. This equation relates the rate of change in the scale factor $a(t)$ to the matter content of the universe. Turning this relation around, we can calculate the change in time per change in scale factor and thus calculate the total age of the universe by integrating this formula. The age t_0 is then given by an expression of the form:

$$t_0 = \frac{1}{H_0} F(\Omega_r, \Omega_m, \Omega_\Lambda, \ldots)$$

where H_0 is the Hubble parameter and the function F depends only on the fractional contribution to the universe's energy content that comes from various components. The first observation that one can make from this formula is that it is the Hubble parameter that controls that age of the universe, with a correction arising from the matter and energy content. So a rough estimate of the age of the universe comes from the Hubble time, the inverse of the Hubble parameter. With a value for H_0 around 69 km/s/Mpc, the Hubble time evaluates to $1/H_0 = 14.5$ billion years.

To get a more accurate number, the correction factor F must be computed. In general this must be done numerically, and the results for a range of cosmological parameter values are shown in the figure. For the Planck values (Ω_m, Ω_Λ) = (0.3086, 0.6914), shown by the box in the upper left corner of the figure, this correction factor is about F = 0.956. For a flat universe without any cosmological constant, shown by the star in the lower right corner, $F = 2/3$ is much smaller and thus the universe is younger for a fixed value of the Hubble parameter. To make this figure, Ω_r is held constant (roughly equivalent to holding the CMB temperature constant) and the curvature density parameter is fixed by the value of the other three.

Apart from the Planck satellite, the Wilkinson Microwave Anisotropy Probe (WMAP) was instrumental in establishing an accurate age of the universe, though other measurements must be folded in to gain an accurate number. CMB measurements are very good at constraining the matter content Ω_m and curvature parameter Ω_k. It is not as sensitive to Ω_Λ directly, partly because the cosmological constant becomes important only at low redshift. The most accurate determinations of the Hubble parameter H_0 come from Type Ia supernovae. Combining these measurements leads to the generally accepted value for the age of the universe quoted above.

The cosmological constant makes the universe "older" for fixed values of the other parameters. This is significant, since before the cosmological constant became generally accepted, the Big Bang model had difficulty explaining why globular clusters in the Milky Way appeared to be far older than the age of the universe as calculated from the Hubble parameter and a matter-only universe. Introducing the cosmological constant allows the universe to be older than these clusters, as well as explaining other features that the matter-only cosmological model could not.

WMAP

NASA's Wilkinson Microwave Anisotropy Probe (WMAP) project's nine-year data release in 2012 estimated the age of the universe to be (13.772±0.059) × 10^9 years (13.772 billion years, with an uncertainty of plus or minus 59 million years).

However, this age is based on the assumption that the project's underlying model is correct; other methods of estimating the age of the universe could give different ages. Assuming an extra background of relativistic particles, for example, can enlarge the error bars of the WMAP constraint by one order of magnitude.

This measurement is made by using the location of the first acoustic peak in the microwave background power spectrum to determine the size of the decoupling surface (size of the universe at the time of recombination). The light travel time to this surface (depending on the geometry used) yields a reliable age for the universe. Assuming the validity of the models used to determine this age, the residual accuracy yields a margin of error near one percent.

Planck

In 2015, the Planck Collaboration estimated the age of the universe to be 13.813 ± 0.038 billion years, slightly higher but within the uncertainties of the earlier number derived from the WMAP data. By combining the Planck data with external data, the best combined estimate of the age of the universe is (13.799±0.021) × 10^9 years old.

In the table below, figures are within 68% confidence limits for the base ΛCDM model.

Legend:

- TT, TE, EE: Planck Cosmic microwave background (CMB) power spectra.
- lowP: Planck polarization data in the low-ℓ likelihood.
- lensing: CMB lensing reconstruction.
- ext: External data (BAO+JLA+H0). BAO: Baryon acoustic oscillations, JLA: Joint Light-curve Analysis, H0: Hubble constant.

Cosmological parameters from 2015 Planck results							
Parameter	Symbol	TT + lowP	TT+lowP +lensing	TT + lowP +lensing+ext	TT,TE, EE+lowP	TT,TE,EE + lowP +lensing	TT,TE, EE+lowP +lensing+ext
Age of the universe (Ga)	t_0	13.813 ± 0.038	13.799 ± 0.038	13.796 ± 0.029	13.813 ± 0.026	13.807 ± 0.026	13.799 ± 0.021
Hubble constant ($\frac{km}{Mpc}$)	H_0	67.31 ± 0.96	67.81 ± 0.92	67.90 ± 0.55	67.27 ± 0.66	67.51 ± 0.64	67.74 ± 0.46

Assumption of Strong Priors

Calculating the age of the universe is accurate only if the assumptions built into the models being used to estimate it are also accurate. This is referred to as strong priors and essentially involves stripping the potential errors in other parts of the model to render the accuracy of actual observational data directly into the concluded result. Although this is not a valid procedure in all contexts, the age given is thus accurate to the specified error (since this error represents the error in the instrument used to gather the raw data input into the model).

The age of the universe based on the best fit to Planck 2015 data alone is 13.813 ± 0.038 billion years (the estimate of 13.799 ± 0.021 billion years uses Gaussian priors based on earlier estimates from other studies to determine the combined uncertainty). This number represents an accurate "direct" measurement of the age of the universe (other methods typically involve Hubble's law and the age of the oldest stars in globular clusters, etc.). It is possible to use different methods for determining the same parameter (in this case – the age of the universe) and arrive at different answers with no overlap in the "errors". To best avoid the problem, it is common to show two sets of uncertainties; one related to the actual measurement and the other related to the systematic errors of the model being used.

An important component to the analysis of data used to determine the age of the universe (e.g. from Planck) therefore is to use a Bayesian statistical analysis, which normalizes the results based upon the priors (i.e. the model). This quantifies any uncertainty in the accuracy of a measurement due to a particular model used.

Cosmological Horizon

A cosmological horizon is a measure of the distance from which one could possibly retrieve information. This observable constraint is due to various properties of general relativity, the expanding universe, and the physics of Big Bang cosmology. Cosmological horizons set the size and scale of the observable universe.

Particle Horizon

The particle horizon (also called the cosmological horizon, the comoving horizon, or the cosmic light horizon) is the maximum distance from which light from particles could have traveled to the observer in the age of the universe. It represents the boundary between the observable and the unobservable regions of the universe, so its distance at the present epoch defines the size of the observable universe. Due to the expansion of the universe it is not simply the age of the universe times the speed of light, as in the Hubble horizon, but rather the speed of light multiplied by the conformal time. The existence, properties, and significance of a cosmological horizon depend on the particular cosmological model.

In terms of comoving distance, the particle horizon is equal to the conformal time that has passed since the Big Bang, times the speed of light. In general, the conformal time at a certain time is given in terms of the scale factor a by:

$$\eta(t) = \int_0^t \frac{dt'}{a(t')}$$

or,

$$\eta(a) = \int_0^a \frac{1}{a'H(a')} \frac{da'}{a'}.$$

The particle horizon is the boundary between two regions at a point at a given time: one region defined by events that have already been observed by an observer, and the other by events which cannot be observed *at that time*. It represents the furthest distance from which we can retrieve information from the past, and so defines the observable universe.

Hubble Horizon

Hubble radius, Hubble sphere, Hubble volume, or Hubble horizon is a conceptual horizon defining the boundary between particles that are moving slower and faster than the speed of light relative to an observer at one given time. Note that this does not mean the particle is unobservable, the light from the past is reaching and will continue to reach the observer for a while. Also, more importantly, in the current expansion models light emitted from the Hubble radius will reach us in a finite amount of time. It is a common misconception that light from the Hubble radius can never reach us. In models assuming decreasing H with time (some cases of Friedmann universe), while particles on the Hubble radius recede from us with the speed of light, the Hubble radius gets larger over time, so light emitted towards us from a particle on the Hubble radius will be inside the Hubble radius some time later. In such models, only light emitted from the cosmic event horizon or further will never reach us in a finite amount of time.

The Hubble velocity of an object is given by Hubble's law:

$$v = xH.$$

Replacing v with speed of light c and solving for proper distance x we obtain the radius of Hubble sphere as:

$$r_{HS}(t) = \frac{c}{H(t)}.$$

In an ever-accelerating universe, if two particles are separated by a distance greater than the Hubble radius, they cannot talk to each other from now on (as they are now, not as they have been in the past), However, if they are outside of each other's particle horizon, they could have never communicated. Depending on the form of expansion of the universe, they may be able to exchange information in the future. Today:

$$r_{HS}(t_0)=\frac{c}{H_0},$$

yielding a Hubble horizon of some 4.1 gigaparsecs. This horizon is not really a physical size, but it is often used as useful length scale as most physical sizes in cosmology can be written in terms of those factors.

One can also define a comoving Hubble horizon by simply dividing the Hubble radius by the scale factor:

$$r_{HS,comoving}(t)=\frac{c}{a(t)H(t)}.$$

Event Horizon

The particle horizon differs from the cosmic event horizon, in that the particle horizon represents the largest comoving distance from which light could have reached the observer by a specific time, while the event horizon is the largest comoving distance from which light emitted now can *ever* reach the observer in the future. The current distance to our cosmic event horizon is about five gigaparsecs (16 billion light-years), well within our observable range given by the particle horizon.

In general, the proper distance to the event horizon at time t is given by:

$$d_e(t)=a(t)\int_t^{t_{max}}\frac{cdt'}{a(t')}$$

where t_{max} is the time-coordinate of the end of the universe, which would be infinite in the case of a universe that expands forever.

For our case, assuming that dark energy is due to a cosmological constant, $d_e(t_0)<\infty$.

Future Horizon

In an accelerating universe, there are events which will be unobservable as $t\to\infty$ as signals from future events become redshifted to arbitrarily long wavelengths in the exponentially expanding de Sitter space. This sets a limit on the farthest distance that we can possibly see as measured in units of proper distance today. Or, more precisely, there are events that are spatially separated for a certain frame of reference happening simultaneously with the event occurring right now for which no signal will ever reach us, even though we can observe events that occurred at the same location in space that happened in the distant past. While we will continue to receive signals from this location in space, even if we wait an infinite amount of time, a signal that left from that location today will never reach us. Additionally, the signals coming from that location will have less and less energy and be less and less frequent until the location, for all practical purposes, becomes unobservable. In a universe that is dominated by dark energy which is undergoing an exponential expansion of the scale factor, all objects that are gravitationally unbound with respect to the Milky Way will become unobservable, in a futuristic version of Kapteyn's universe.

Practical Horizons

While not technically "horizons" in the sense of an impossibility for observations due to relativity or cosmological solutions, there are practical horizons which include the optical horizon, set at the surface of last scattering. This is the farthest distance that any photon can freely stream. Similarly, there is a "neutrino horizon" set for the farthest distance a neutrino can freely stream and a gravitational wave horizon at the farthest distance that gravitational waves can freely stream. The latter is predicted to be a direct probe of the end of cosmic inflation.

Timeline of the Big Bang

Since the Big Bang, 13.7 billion years ago, the universe has passed through many different phases or epochs. Due to the extreme conditions and the violence of its very early stages, it arguably saw more activity and change during the first second than in all the billions of years since.

From our current understanding of how the Big Bang might have progressed, taking into account theories about inflation, Grand Unification, etc., we can put together an approximate timeline as follows:

- Planck Epoch (or Planck Era), from zero to approximately 10^{-43} seconds (1 Planck Time):

  This is the closest that current physics can get to the absolute beginning of time, and very little can be known about this period. General relativity proposes a gravitational singularity before this time (although even that may break down due to quantum effects), and it is hypothesized that the four fundamental forces (electromagnetism, weak nuclear force, strong nuclear force and gravity) all have the same strength, and are possibly even unified into one fundamental force, held together by a perfect symmetry which some have likened to a sharpened pencil standing on its point (i.e. too symmetrical to last). At this point, the universe spans a region of only 10^{-35}

meters (1 Planck Length), and has a temperature of over 1032°C (the Planck Temperature).

- Grand Unification Epoch, from 10^{-43} seconds to 10^{-36} seconds:

  The force of gravity separates from the other fundamental forces (which remain unified), and the earliest elementary particles (and antiparticles) begin to be created.

- Inflationary Epoch, from 10^{-36} seconds to 10^{-32} seconds:

  Triggered by the separation of the strong nuclear force, the universe undergoes an extremely rapid exponential expansion, known as cosmic inflation. The linear dimensions of the early universe increases during this period of a tiny fraction of a second by a factor of at least 1026 to around 10 centimeters (about the size of a grapefruit). The elementary particles remaining from the Grand Unification Epoch (a hot, dense quark-gluon plasma, sometimes known as "quark soup") become distributed very thinly across the universe.

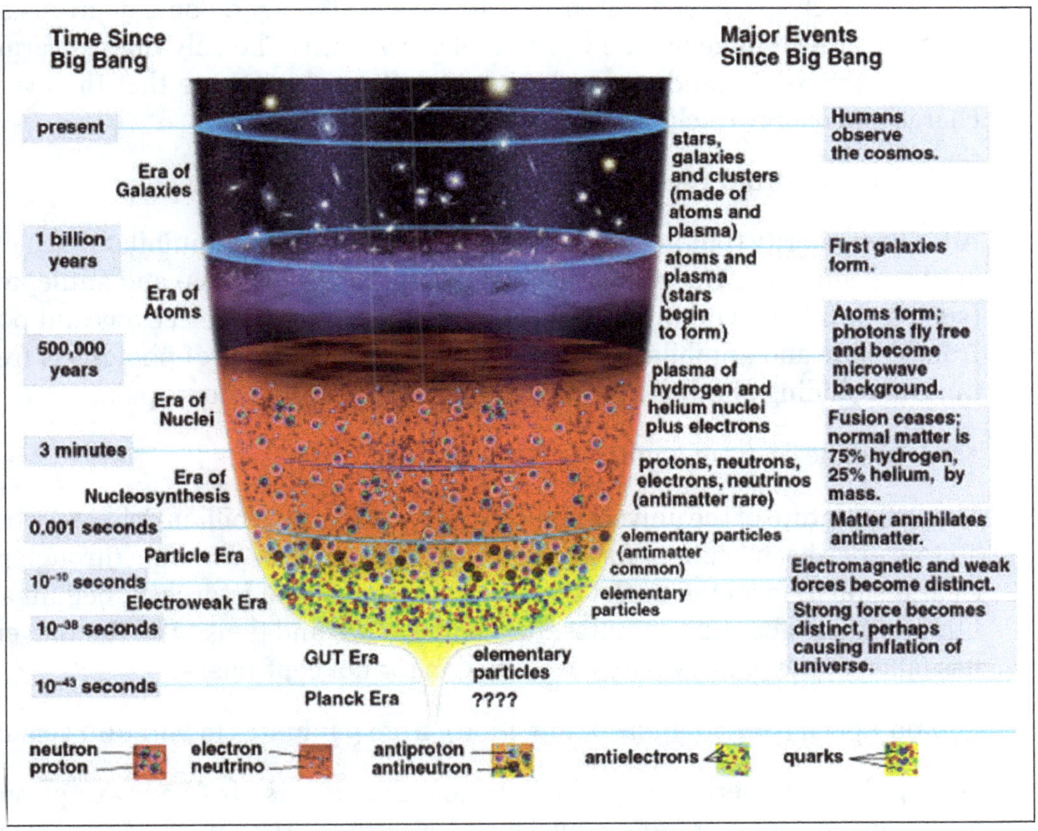

Timeline and major events since the Big Bang.

- Electroweak Epoch, from 10^{-36} seconds to 10^{-12} seconds:

  As the strong nuclear force separates from the other two, particle interactions create large numbers of exotic particles, including W and Z bosons and Higgs bosons (the Higgs field slows particles down and confers mass on them, allowing a universe made entirely out of radiation to support things that have mass).

- Quark Epoch, from 10^{-12} seconds to 10^{-6} seconds:

  Quarks, electrons and neutrinos form in large numbers as the universe cools off to below 10 quadrillion degrees, and the four fundamental forces assume their present forms. Quarks and antiquarks annihilate each other upon contact, but, in a process known as baryogenesis, a surplus of quarks (about one for every billion pairs) survives, which will ultimately combine to form matter.

- Hadron Epoch, from 10^{-6} seconds to 1 second:

  The temperature of the universe cools to about a trillion degrees, cool enough to allow quarks to combine to form hadrons (like protons and neutrons). Electrons colliding with protons in the extreme conditions of the Hadron Epoch fuse to form neutrons and give off massless neutrinos, which continue to travel freely through space today, at or near to the speed of light. Some neutrons and neutrinos re-combine into new proton-electron pairs. The only rules governing all this apparently random combining and re-combining are that the overall charge and energy (including mass-energy) be conserved.

- Lepton Epoch, from 1 second to 3 minutes:

  After the majority (but not all) of hadrons and antihadrons annihilate each other at the end of the Hadron Epoch, leptons (such as electrons) and antileptons (such as positrons) dominate the mass of the universe. As electrons and positrons collide and annihilate each other, energy in the form of photons is freed up, and colliding photons in turn create more electron-positron pairs.

- Nucleosynthesis, from 3 minutes to 20 minutes:

  The temperature of the universe falls to the point (about a billion degrees) where atomic nuclei can begin to form as protons and neutrons combine through nuclear fusion to form the nuclei of the simple elements of hydrogen, helium and lithium. After about 20 minutes, the temperature and density of the universe has fallen to the point where nuclear fusion cannot continue.

- Photon Epoch (or Radiation Domination), from 3 minutes to 240,000 years:

  During this long period of gradual cooling, the universe is filled with plasma, a hot, opaque soup of atomic nuclei and electrons. After most of the leptons and antileptons had annihilated each other at the end of the Lepton Epoch, the energy of the universe is dominated by photons, which continue to interact frequently with the charged protons, electrons and nuclei.

- Recombination/Decoupling, from 240,000 to 300,000 years:

  As the temperature of the universe falls to around 3,000 degrees (about the same heat as the surface of the Sun) and its density also continues to fall, ionized hydrogen and helium atoms capture electrons (known as "recombination"), thus neutralizing their electric charge. With the electrons now bound to atoms, the universe finally becomes transparent to light, making this the earliest epoch observable today. It also releases the photons in the universe which have up till this time been interacting with electrons and protons in an opaque photon-baryon fluid (known as "decoupling"), and these photons (the same ones we see in today's cosmic background radiation) can now travel freely. By the end of this period, the universe consists of a fog of about 75% hydrogen and 25% helium, with just traces of lithium.

- Dark Age (or Dark Era), from 300,000 to 150 million years:

  The period after the formation of the first atoms and before the first stars is sometimes referred to as the Dark Age. Although photons exist, the universe at this time is literally dark, with no stars having formed to give off light. With only very diffuse matter remaining, activity in the universe has tailed off dramatically, with very low energy levels and very large time scales. Little of note happens during this period, and the universe is dominated by mysterious "dark matter".

- Reionization, 150 million to 1 billion years:

  The first quasars form from gravitational collapse, and the intense radiation they emit reionizes the surrounding universe, the second of two major phase changes of hydrogen gas in the universe (the first being the Recombination period). From this point on, most of the universe goes from being neutral back to being composed of ionized plasma.

- Star and Galaxy Formation, 300 - 500 million years onwards:

  Gravity amplifies slight irregularities in the density of the primordial gas and pockets of gas become more and more dense, even as the universe continues to expand rapidly. These small, dense clouds of cosmic gas start to collapse under their own gravity, becoming hot enough to trigger nuclear fusion reactions between hydrogen atoms, creating the very first stars.

  The first stars are short-lived supermassive stars, a hundred or so times the mass of our Sun, known as Population III (or "metal-free") stars. Eventually Population II and then Population I stars also begin to form from the material from previous rounds of star-making. Larger stars burn out quickly and explode in massive supernova events, their ashes going to form subsequent generations of stars. Large volumes of matter collapse to form galaxies and gravitational attraction pulls galaxies towards each other to form groups, clusters and superclusters.

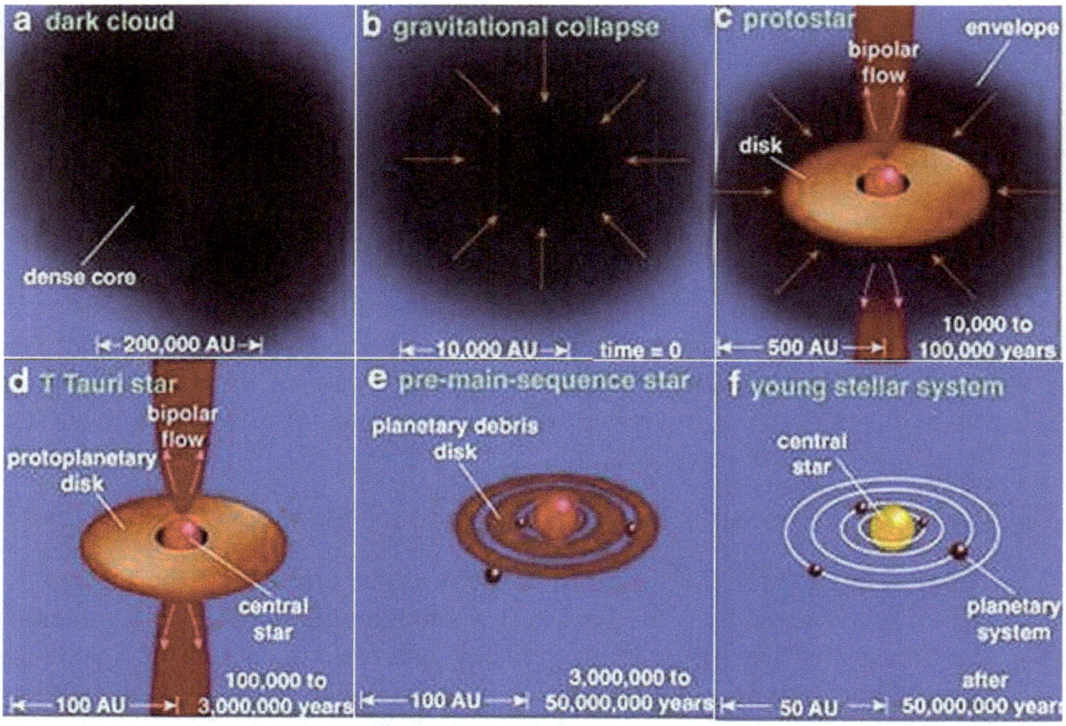

The process of star formation.

- Solar System Formation, 8.5 - 9 billion years:

  Our Sun is a late-generation star, incorporating the debris from many generations of earlier stars, and it and the Solar System around it form roughly 4.5 to 5 billion years ago (8.5 to 9 billion years after the Big Bang).

- Today, 13.7 billion years:

  The expansion of the universe and recycling of star materials into new stars continues.

References

- Big-bang-theory-25126: space.com, Retrieved 24 July, 2020
- Maeder, Andre (2017). "An Alternative to the ΛCDM Model: The Case of Scale Invariance". The Astrophysical Journal. 834 (2): 194. arXiv:1701.03964. Bibcode:2017ApJ...834..194M. doi:10.3847/1538-4357/834/2/194. ISSN 0004-637X
- Margalef-Bentabol, Berta; Margalef-Bentabol, Juan; Cepa, Jordi (8 February 2013). "Evolution of the cosmological horizons in a universe with countably infinitely many state equations". Journal of Cosmology and Astroparticle Physics. 015. 2013 (2): 015. arXiv:1302.2186. Bibcode:2013JCAP...02..015M. doi:10.1088/1475-7516/2013/02/015
- Sandage, A. R. (1958). "Current Problems in the Extragalactic Distance Scale". The Astrophysical Journal. 127 (3): 513–526. Bibcode:1958ApJ...127..513S. doi:10.1086/146483
- Topics-bigbang-timeline: physicsoftheuniverse.com, Retrieved 24 August, 2020

5
Galaxies

A gravitationally bound system of stars, stellar remnants, interstellar gas, dust, and dark matter is termed as a galaxy. Some of the galaxies are Milky Way, Andromeda galaxy, elliptical galaxy, spiral galaxy, starburst galaxy, etc. This chapter closely examines these galaxies to provide an extensive understanding of the subject.

Galaxies are sprawling systems of dust, gas, dark matter, and anywhere from a million to a trillion stars that are held together by gravity. Nearly all large galaxies are thought to also contain supermassive black holes at their centers. In our own galaxy, the Milky Way, the sun is just one of about 100 to 400 billion stars that spin around Sagittarius A*, a supermassive black hole that contains as much mass as four million suns.

The deeper we look into the cosmos, the more galaxies we see. One 2016 study estimated that the observable universe contains two trillion—or two million million—galaxies. Some of those distant systems are similar to our own Milky Way galaxy, while others are quite different.

Types of Galaxies

Before the 20th century, we didn't know that galaxies other than the Milky Way existed; earlier astronomers had classified them as as "nebulae," since they looked like fuzzy clouds. But in the 1920s, astronomer Edwin Hubble showed that the Andromeda "nebula" was a galaxy in its own right. Since it is so far from us, it takes light from Andromeda more than 2.5 million years to bridge the gap. Despite the immense distance, Andromeda is the closest large galaxy to our Milky Way, and it's bright enough in the night sky that it's visible to the naked eye in the Northern Hemisphere.

In 1936, Hubble debuted a way to classify galaxies, grouping them into four main types: spiral galaxies, lenticular galaxies, elliptical galaxies, and irregular galaxies.

More than two-thirds of all observed galaxies are spiral galaxies. A spiral galaxy has a flat, spinning disk with a central bulge surrounded by spiral arms. That spinning motion, at speeds of hundreds of kilometers a second, may cause matter in the disk to take on a distinctive spiral shape, like a cosmic pinwheel. Our Milky Way, like other spiral galaxies, has a linear, starry bar at its center.

Elliptical galaxies are shaped as their name suggests: They are generally round but can stretch longer along one axis than along the other, so much so that some take on a cigar-like appearance. The universe's largest-known galaxies—giant elliptical galaxies—can contain up to a trillion stars and span two million light-years across. Elliptical galaxies may also be small, in which case they are called dwarf elliptical galaxies.

Elliptical galaxies contain many older stars, but little dust and other interstellar matter. Their stars orbit the galactic center, like those in the disks of spiral galaxies, but they do so in more random directions. Few new stars are known to form in elliptical galaxies. They are common in galaxy clusters.

Lenticular galaxies, such as the iconic Sombrero Galaxy, sit between elliptical and spiral galaxies. They're called "lenticular" because they resemble lenses: Like spiral galaxies, they have a thin, rotating disk of stars and a central bulge, but they don't have spiral arms. Like elliptical galaxies, they have little dust and interstellar matter, and they seem to form more often in densely populated regions of space.

Galaxies that are not spiral, lenticular, or elliptical are called irregular galaxies. Irregular galaxies—such as the Large and Small Magellanic Clouds that flank our Milky Way—appear misshapen and lack a distinct form, often because they are within the gravitational influence of other galaxies close by. They are full of gas and dust, which makes them great nurseries for forming new stars.

Galactic Clusters and Mergers

Some galaxies occur alone or in pairs, but they are more often parts of larger associations known as groups, clusters, and superclusters. Our Milky Way, for instance, is in the Local Group, a galaxy group about 10 million light-years across that also includes the Andromeda galaxy and its satellites. The Local Group and its neighbor galaxy cluster, the Virgo Cluster, both lie within the larger Virgo Supercluster, a concentration of galaxies that stretches about 100 million light-years across. The Virgo Supercluster, in turn, is a limb of Laniakea, an even bigger supercluster of 100,000 galaxies that astronomers defined in 2014.

Galaxies in clusters often interact and even merge together in a dynamic cosmic dance of interacting gravity. When two galaxies collide and intermingle, gases can flow towards the galactic center, which can trigger phenomena like rapid star formation. Our own Milky Way will merge with the Andromeda galaxy in about 4.5 billion years.

Because elliptical galaxies contain older stars and less gas than spiral galaxies, it seems that the galaxy types represent part of a natural evolution: As spiral galaxies age, interact, and merge, they lose their familiar shapes and become elliptical galaxies. But astronomers are still working out the specifics, such as why elliptical galaxies follow certain patterns in brightness, size, and chemical composition.

Milky Way

The Milky Way[a] is the galaxy that contains our Solar System, with the name describing the galaxy's appearance from Earth: a hazy band of light seen in the night sky formed from stars that cannot be individually distinguished by the naked eye. From Earth, the Milky Way appears as a band because its disk-shaped structure is viewed from within. Galileo Galilei first resolved the band of light into individual stars with his telescope in 1610. Until the early 1920s, most astronomers thought that the Milky Way contained all the stars in the Universe. Following the 1920 Great Debate between the astronomers Harlow Shapley and Heber Curtis, observations by Edwin Hubble showed that the Milky Way is just one of many galaxies.

The Milky Way is a barred spiral galaxy with a diameter between 150,000 and 200,000 light-years (ly). It is estimated to contain 100–400 billion stars and more than 100 billion planets. The Solar System is located at a radius of about 27,000 light-years from the Galactic Center, on the inner edge of the Orion Arm, one of the spiral-shaped concentrations of gas and dust. The stars in the innermost 10,000 light-years form a bulge and one or more bars that radiate from the bulge. The galactic center is an intense radio source known as Sagittarius A*, assumed to be a supermassive black hole of 4.100 (± 0.034) million solar masses.

Stars and gases at a wide range of distances from the Galactic Center orbit at approximately 220 kilometers per second. The constant rotation speed contradicts the laws of Keplerian dynamics and suggests that much (about 90%) of the mass of the Milky Way is invisible to telescopes, neither emitting nor absorbing electromagnetic radiation. This conjectural mass has been termed "dark matter". The rotational period is about 240 million years at the radius of the Sun. The Milky Way as a whole is moving at a velocity of approximately 600 km per second with respect to extragalactic frames of reference. The oldest stars in the Milky Way are nearly as old as the Universe itself and thus probably formed shortly after the Dark Ages of the Big Bang.

The Milky Way has several satellite galaxies and is part of the Local Group of galaxies, which form part of the Virgo Supercluster, which is itself a component of the Laniakea Supercluster.

Appearance

The Milky Way is visible from Earth as a hazy band of white light, some 30° wide, arching across the night sky. In night sky observing, although all the individual naked-eye stars in the entire sky are part of the Milky Way, the term "Milky Way" is limited to this band of light. The light originates from the accumulation of unresolved stars and other material located in the direction of the galactic plane. Dark regions within the band, such as the Great Rift and the Coalsack, are areas where interstellar dust blocks light from distant stars. The area of sky that the Milky Way obscures is called the Zone of Avoidance.

A view of the Milky Way toward the constellation Sagittarius (including the Galactic Center), as seen from a dark site with little light pollution (the Black Rock Desert), the bright object on the lower right is Jupiter, just above Antares.

The Milky Way has a relatively low surface brightness. Its visibility can be greatly reduced by background light, such as light pollution or moonlight. The sky needs to be darker than about 20.2 magnitude per square arcsecond in order for the Milky Way to be visible. It should be visible if the limiting magnitude is approximately +5.1 or better and shows a great deal of detail at +6.1. This makes the Milky Way difficult to see from brightly lit urban or suburban areas, but very prominent when viewed from rural areas when the Moon is below the horizon. Maps of artificial night sky brightness show that more than one-third of Earth's population cannot see the Milky Way from their homes due to light pollution.

As viewed from Earth, the visible region of the Milky Way's galactic plane occupies an area of the sky that includes 30 constellations. The Galactic Center lies in the direction of Sagittarius, where the Milky Way is brightest. From Sagittarius, the hazy band of white light appears to pass around to the galactic anticenter in Auriga. The band then continues the rest of the way around the sky, back to Sagittarius, dividing the sky into two roughly equal hemispheres.

The galactic plane is inclined by about 60° to the ecliptic (the plane of Earth's orbit). Relative to the celestial equator, it passes as far north as the constellation of Cassiopeia and as far south as the constellation of Crux, indicating the high inclination of Earth's equatorial plane and the plane of the ecliptic, relative to the galactic plane. The north galactic pole is situated at right ascension $12^h\ 49^m$, declination +27.4° (B1950) near β Comae Berenices, and the south galactic pole is near α Sculptoris. Because of this high inclination, depending on the time of night and year, the arch of the Milky Way may appear relatively low or relatively high in the sky. For observers from latitudes approximately 65° north to 65° south, the Milky Way passes directly overhead twice a day.

The Milky Way arching at a high inclination across the night sky, the bright object is Jupiter in the constellation Sagittarius, and the Magellanic Clouds can be seen on the left; galactic north is downward.

Size and Mass

A HST image of galaxy UGC 12158, which is thought to resemble the Milky Way in appearance.

The Milky Way is the second-largest galaxy in the Local Group, with its stellar disk approximately 100,000 ly (30 kpc) in diameter and, on average, approximately 1,000 ly (0.3 kpc) thick. The Milky Way is approximately 890 billion times the mass of the Sun. To compare the relative physical scale of the Milky Way, if the Solar System out to Neptune were the size of a US quarter (24.3 mm (0.955 in)), the Milky Way would be approximately the size of the contiguous United States. There is a ring-like filament of stars rippling above and below the relatively flat galactic plane, wrapping around the Milky Way at a diameter of 150,000–180,000 light-years (46–55 kpc), which may be part of the Milky Way itself.

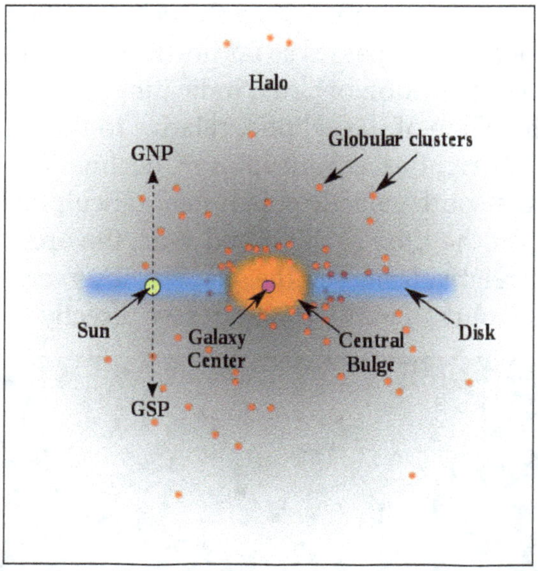

A schematic profile of the Milky Way.

Estimates of the mass of the Milky Way vary, depending upon the method and data used. The low end of the estimate range is 5.8×10[11] solar masses ($M_\odot$), somewhat less than that of the Andromeda Galaxy. Measurements using the Very Long Baseline Array in 2009 found velocities as large as 254 km/s (570,000 mph) for stars at the outer edge of the Milky Way. Because the orbital velocity depends on the total mass inside the orbital radius, this suggests that the Milky Way is more massive, roughly equaling the mass of Andromeda Galaxy at $7\times10^{11}\ M_\odot$ within 160,000 ly (49 kpc) of its center. In 2010, a measurement of the radial velocity of halo stars found that the mass enclosed within 80 kiloparsecs is $7\times10^{11}\ M_\odot$. According to a study published in 2014, the mass of the entire Milky Way is estimated to be $8.5\times10^{11}\ M_\odot$, but this is only half the mass of the Andromeda Galaxy. A recent mass estimate for the Milky Way is $1.29\times10^{12}\ M_\odot$.

Much of the mass of the Milky Way seems to be dark matter, an unknown and invisible form of matter that interacts gravitationally with ordinary matter. A dark matter halo is conjectured to spread out relatively uniformly to a distance beyond one hundred kiloparsecs (kpc) from the Galactic Center. Mathematical models of the Milky Way suggest that the mass of dark matter is $1–1.5\times10^{12}\ M_\odot$. Recent studies indicate a range in mass, as large as $4.5\times10^{12}\ M_\odot$ and as small as $8\times10^{11}\ M_\odot$. The total mass of all the stars in the Milky Way is estimated to be between $4.6\times10^{10}\ M_\odot$ and $6.43\times10^{10}\ M_\odot$. In addition to the stars, there is also interstellar gas, comprising 90% hydrogen and 10% helium by mass, with two thirds of the hydrogen found in the atomic form and the remaining one-third as molecular hydrogen. The mass of the Milky Way's interstellar gas is equal to between 10% and 15% of the total mass of its stars. Interstellar dust accounts for an additional 1% of the total mass of the gas.

In March 2019, astronomers reported that the mass of the Milky Way galaxy is 1.5 trillion solar masses within a radius of about 129,000 light-years, over twice as much as was determined in earlier studies, and suggesting that about 90% of the mass of the galaxy is dark matter.

Contents

The Milky Way contains between 100 and 400 billion stars and at least 100 billion planets. An exact figure would depend on counting the number of very-low-mass stars, which are difficult to detect, especially at distances of more than 300 ly (90 pc) from the Sun. As a comparison, the neighboring Andromeda Galaxy contains an estimated one trillion (10^{12}) stars. Perhaps, the Milky Way may contain ten billion white dwarfs, a billion neutron stars, and a hundred million stellar black holes. Filling the space between the stars is a disk of gas and dust called the interstellar medium. This disk has at least a comparable extent in radius to the stars, whereas the thickness of the gas layer ranges from hundreds of light-years for the colder gas to thousands of light-years for warmer gas.

The disk of stars in the Milky Way does not have a sharp edge beyond which there are no stars. Rather, the concentration of stars decreases with distance from the center of the Milky Way. For reasons that are not understood, beyond a radius of roughly 40,000 ly (13 kpc) from the center, the number of stars per cubic parsec drops much faster with radius. Surrounding the galactic disk is a spherical Galactic Halo of stars and globular clusters that extends farther outward, but is limited in size by the orbits of two Milky Way satellites, the Large and Small Magellanic Clouds, whose closest approach to the Galactic Center is about 180,000 ly (55 kpc). At this distance or beyond, the orbits of most halo objects would be disrupted by the Magellanic Clouds. Hence, such objects would probably be ejected from the vicinity of the Milky Way. The integrated absolute visual magnitude of the Milky Way is estimated to be around −20.9.

360-degree panorama view of the Milky Way, the galactic centre is in the middle of the view, with galactic north up.

Both gravitational microlensing and planetary transit observations indicate that there may be at least as many planets bound to stars as there are stars in the Milky Way, and microlensing measurements indicate that there are more rogue planets not bound to host stars than there are stars. The Milky Way contains at least one planet per star, resulting in 100–400 billion planets, according to a January 2013 study of the five-planet star system Kepler-32 with the Kepler space observatory. A different January 2013 analysis of Kepler data estimated that at least 17 billion Earth-sized exoplanets reside in the Milky Way. On November 4, 2013, astronomers reported, based on *Kepler* space mission data, that there could be as many as 40 billion Earth-sized planets orbiting in the habitable zones of Sun-like stars and red dwarfs within the Milky Way. 11 billion of these estimated planets may be orbiting Sun-like stars. The nearest exoplanet may be 4.2 light-years away, orbiting the red dwarf Proxima Centauri, according to a 2016 study. Such Earth-sized planets may be more numerous than gas giants. Besides exoplanets, "exocomets", comets beyond the Solar System, have also been detected and may be common in the Milky Way.

Structure

The Milky Way consists of a bar-shaped core region surrounded by a warped disk of gas, dust and stars. The mass distribution within the Milky Way closely resembles the type Sbc in the Hubble classification, which represents spiral galaxies with relatively loosely wound arms. Astronomers first began to suspect that the Milky Way is a barred spiral galaxy, rather than an ordinary spiral galaxy, in the 1960s. Their suspicions were confirmed by the Spitzer Space Telescope observations in 2005 that showed the Milky Way's central bar to be larger than previously thought.

Spitzer reveals what cannot be seen in visible light: cooler stars (blue), heated dust (reddish hue), and Sgr A* as bright white spot in the middle.

Bright X-ray flares from Sagittarius A*, location of the supermassive black hole at the center of the Milky Way.

Galactic Quadrants

A galactic quadrant, or quadrant of the Milky Way, refers to one of four circular sectors in the division of the Milky Way. In astronomical practice, the delineation of the galactic quadrants is based upon the galactic coordinate system, which places the Sun as the origin of the mapping system.

Quadrants are described using ordinals—for example, "1st galactic quadrant", "second galactic quadrant", or "third quadrant of the Milky Way". Viewing from the north galactic pole with 0 degrees (°) as the ray that runs starting from the Sun and through the Galactic Center, the quadrants are as follows:

- 1st galactic quadrant – $0° \leq$ longitude (ℓ) $\leq 90°$.
- 2nd galactic quadrant – $90° \leq \ell \leq 180°$.
- 3rd galactic quadrant – $180° \leq \ell \leq 270°$.
- 4th galactic quadrant – $270° \leq \ell \leq 360°$ (0°).

Galactic Center

The Sun is 25,000–28,000 ly (7.7–8.6 kpc) from the Galactic Center. This value is estimated using geometric-based methods or by measuring selected astronomical objects that serve as standard candles, with different techniques yielding various values within this approximate range. In the inner few kpc (around 10,000 light-years radius) is a dense concentration of mostly old stars in a roughly spheroidal shape called the bulge. It has been proposed that the Milky Way lacks a bulge formed due to a collision and merger between previous galaxies, and that instead it only has a *pseudobulge* formed by its central bar. However, confusion in the literature between the (peanut shell)-shaped structure created by instabilities in the bar, versus a possible bulge with an expected half-light radius of 0.5 kpc, abound.

The Galactic Center is marked by an intense radio source named Sagittarius A*. The motion of material around the center indicates that Sagittarius A* harbors a massive, compact object. This concentration of mass is best explained as a supermassive black hole (SMBH) with an estimated mass of 4.1–4.5 million times the mass of the Sun. The rate of accretion of the SMBH is consistent with an inactive galactic nucleus, being estimated at around $1 \times 10^{-5}\ M_{\odot}$ per year. Observations indicate that there are SMBHs located near the center of most normal galaxies.

The nature of the Milky Way's bar is actively debated, with estimates for its half-length and orientation spanning from 1 to 5 kpc (3,000–16,000 ly) and 10–50 degrees relative to the line of sight from Earth to the Galactic Center. Certain authors advocate that the Milky Way features two distinct bars, one nestled within the other. However, RR Lyrae variables do not trace a prominent Galactic bar. The bar may be surrounded by a ring called the "5-kpc ring" that contains a large fraction of the molecular hydrogen present in the Milky Way, as well as most of the Milky Way's star formation activity. Viewed from the Andromeda Galaxy, it would be the brightest feature of the Milky Way. X-ray emission from the core is aligned with the massive stars surrounding the central bar and the Galactic ridge.

Illustration of the two gigantic X-ray/gamma-ray bubbles (blue-violet) of the Milky Way (center).

In 2010, two gigantic spherical bubbles of high energy emission were detected to the north and the south of the Milky Way core, using data from the Fermi Gamma-ray Space Telescope. The diameter of each of the bubbles is about 25,000 light-years (7.7 kpc); they stretch up to Grus and to Virgo on the night-sky of the southern hemisphere. Subsequently, observations with the Parkes Telescope at radio frequencies identified polarized emission that is associated with the Fermi bubbles. These observations are best interpreted as a magnetized outflow driven by star formation in the central 640 ly (200 pc) of the Milky Way.

Later, on January 5, 2015, NASA reported observing an X-ray flare 400 times brighter than usual, a record-breaker, from Sagittarius A*. The unusual event may have been caused by the breaking apart of an asteroid falling into the black hole or by the entanglement of magnetic field lines within gas flowing into Sagittarius A*.

Spiral Arms

Outside the gravitational influence of the Galactic bar, the structure of the interstellar medium and stars in the disk of the Milky Way is organized into four spiral arms. Spiral arms typically contain a higher density of interstellar gas and dust than the Galactic average as well as a greater concentration of star formation, as traced by H II regions and molecular clouds.

The Milky Way's spiral structure is uncertain, and there is currently no consensus on the nature of the Milky Way's spiral arms. Perfect logarithmic spiral patterns only crudely describe features near the Sun, because galaxies commonly have arms that branch, merge, twist unexpectedly, and feature a degree of irregularity. The possible scenario of the Sun within a spur / Local arm emphasizes that point and indicates that such features are probably not unique, and exist elsewhere in the Milky Way. Estimates of the pitch angle of the arms range from about 7° to 25°. There are thought to be four spiral arms that all start near the Milky Way's center. These are named as follows, with the positions of the arms shown in the image at right:

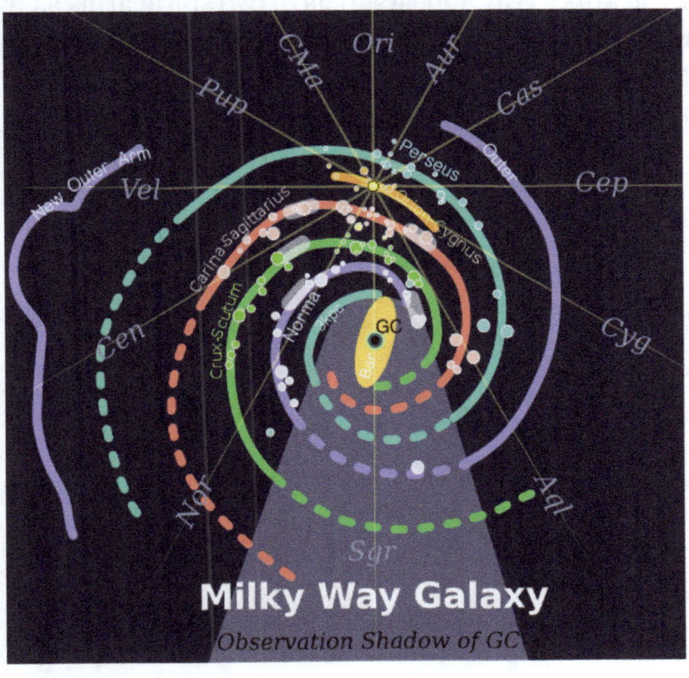

Observed (normal lines) and extrapolated (dotted lines) structure of the spiral arms of the Milky Way, viewed from "north" of the galaxy – stars generally move clockwise in this view. The gray lines radiating from the Sun's position (upper center) list the three-letter abbreviations of the corresponding constellations.

Color	Arm(s)
Cyan	Near 3 kpc Arm and Perseus Arm
Purple	Norma and Outer arm (Along with extension discovered in 2004)
Green	Scutum–Centaurus Arm
Pink	Carina-Sagittarius Arm
There are at least two smaller arms or spurs, including:	
Orange	Orion–Cygnus Arm (which contains the Sun and Solar System)

Two spiral arms, the Scutum–Centaurus arm and the Carina–Sagittarius arm, have tangent points inside the Sun's orbit about the center of the Milky Way. If these arms contain an overdensity of stars compared to the average density of stars in the Galactic disk, it would be detectable by counting the stars near the tangent point. Two surveys of near-infrared light, which is sensitive primarily to red giants and not affected by dust extinction, detected the predicted overabundance in the Scutum–Centaurus arm but not in the Carina–Sagittarius arm: the Scutum–Centaurus Arm contains approximately 30% more red giants than would be expected in the absence of a spiral arm. This observation suggests that the Milky Way possesses only two major stellar arms: the Perseus arm and the Scutum–Centaurus arm. The rest of the arms contain excess gas but not excess old stars. In December 2013, astronomers found that the distribution of young stars and star-forming regions matches the four-arm spiral description of the Milky Way. Thus, the Milky Way appears to have two spiral arms as traced by old stars and four spiral arms as traced by gas and young stars. The explanation for this apparent discrepancy is unclear.

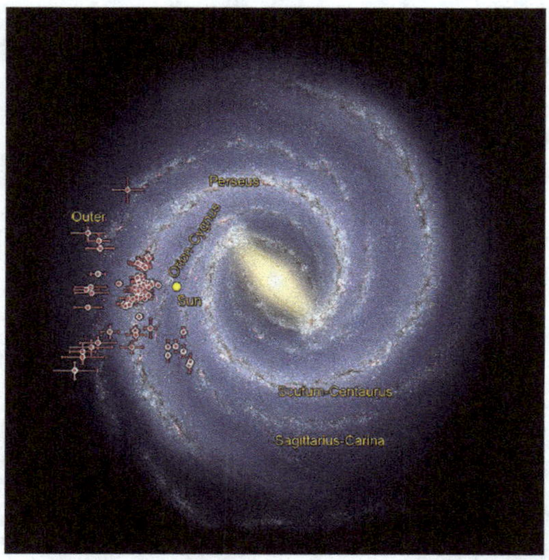

Clusters detected by WISE used to trace the Milky Way's spiral arms.

The Near 3 kpc Arm (also called Expanding 3 kpc Arm or simply 3 kpc Arm) was discovered in the 1950s by astronomer van Woerden and collaborators through 21-centimeter radio measurements of HI (atomic hydrogen). It was found to be expanding away from the central bulge at more than 50 km/s. It is located in the fourth galactic quadrant at a distance of about 5.2 kpc from the Sun and 3.3 kpc from the Galactic Center. The Far 3 kpc Arm was discovered in 2008 by astronomer Tom Dame (Harvard–Smithsonian CfA). It is located in the first galactic quadrant at a distance of 3 kpc (about 10,000 ly) from the Galactic Center.

A simulation published in 2011 suggested that the Milky Way may have obtained its spiral arm structure as a result of repeated collisions with the Sagittarius Dwarf Elliptical Galaxy.

It has been suggested that the Milky Way contains two different spiral patterns: an inner one, formed by the Sagittarius arm, that rotates fast and an outer one, formed by the Carina and Perseus arms, whose rotation velocity is slower and whose arms are tightly wound. In this scenario, suggested by numerical simulations of the dynamics of the different spiral arms, the outer pattern would form an outer pseudoring, and the two patterns would be connected by the Cygnus arm.

The long filamentary molecular cloud dubbed "Nessie" probably forms a dense "spine" of the Scutum–Centarus Arm.

Outside of the major spiral arms is the Monoceros Ring (or Outer Ring), a ring of gas and stars torn from other galaxies billions of years ago. However, several members of the scientific community recently restated their position affirming the Monoceros structure is nothing more than an over-density produced by the flared and warped thick disk of the Milky Way. The structure of the Milky Way's disk is warped along an "S" curve.

Halo

The Galactic disk is surrounded by a spheroidal halo of old stars and globular clusters, of which 90% lie within 100,000 light-years (30 kpc) of the Galactic Center. However, a few globular clusters have been found farther, such as PAL 4 and AM1 at more than 200,000 light-years from the Galactic Center. About 40% of the Milky Way's clusters are on retrograde orbits, which means they move in the opposite direction from the Milky Way rotation. The globular clusters can follow rosette orbits about the Milky Way, in contrast to the elliptical orbit of a planet around a star.

Although the disk contains dust that obscures the view in some wavelengths, the halo component does not. Active star formation takes place in the disk (especially in the spiral arms, which represent areas of high density), but does not take place in the halo, as there is little cool gas to collapse into stars. Open clusters are also located primarily in the disk.

Discoveries in the early 21st century have added dimension to the knowledge of the Milky Way's structure. With the discovery that the disk of the Andromeda Galaxy (M31) extends much farther than previously thought, the possibility of the disk of the Milky Way extending farther is apparent, and this is supported by evidence from the discovery of the Outer Arm extension of the Cygnus Arm and of a similar extension of the Scutum–Centaurus Arm. With the discovery of the Sagittarius Dwarf Elliptical Galaxy came the discovery of a ribbon of galactic debris as the polar orbit of the dwarf and its interaction with the Milky Way tears it apart. Similarly, with the discovery of the Canis Major Dwarf Galaxy, it was found that a ring of galactic debris from its interaction with the Milky Way encircles the Galactic disk.

The Sloan Digital Sky Survey of the northern sky shows a huge and diffuse structure (spread out across an area around 5,000 times the size of a full moon) within the Milky Way that does not seem to fit within current models. The collection of stars rises close to perpendicular to the plane of the spiral arms of the Milky Way. The proposed likely interpretation is that a dwarf galaxy is merging with the Milky Way. This galaxy is tentatively named the Virgo Stellar Stream and is found in the direction of Virgo about 30,000 light-years (9 kpc) away.

Gaseous Halo

In addition to the stellar halo, the Chandra X-ray Observatory, XMM-Newton, and *Suzaku* have provided evidence that there is a gaseous halo with a large amount of hot gas. The halo extends for hundreds of thousand of light-years, much farther than the stellar halo and close to the distance of the Large and Small Magellanic Clouds. The mass of this hot halo is nearly equivalent to the mass of the Milky Way itself. The temperature of this halo gas is between 1 and 2.5 million K (1.8 and 4.5 million °F).

Observations of distant galaxies indicate that the Universe had about one-sixth as much baryonic (ordinary) matter as dark matter when it was just a few billion years old. However, only about half of those baryons are accounted for in the modern Universe based on observations of nearby galaxies like the Milky Way. If the finding that the mass of the halo is comparable to the mass of the Milky Way is confirmed, it could be the identity of the missing baryons around the Milky Way.

Sun's Location and Neighborhood

The Sun is near the inner rim of the Orion Arm, within the Local Fluff of the Local Bubble, and in the Gould Belt. Based upon studies of stellar orbits around Sgr A* by Gillessen and associates, the Sun lies at an estimated distance of 27.14 ± 0.46 kly (8.32 ± 0.14 kpc) from the Galactic Center. Boehle and associates found a smaller value of 25.64 ± 0.46 kly (7.86 ± 0.14 kpc), also using a star orbit analysis. The Sun is currently 5–30 parsecs (16–98 ly) above, or north of, the central plane of the Galactic disk. The distance between the local arm and the next arm out, the Perseus Arm, is about 2,000 parsecs (6,500 ly). The Sun, and thus the Solar System, is located in the Milky Way's galactic habitable zone.

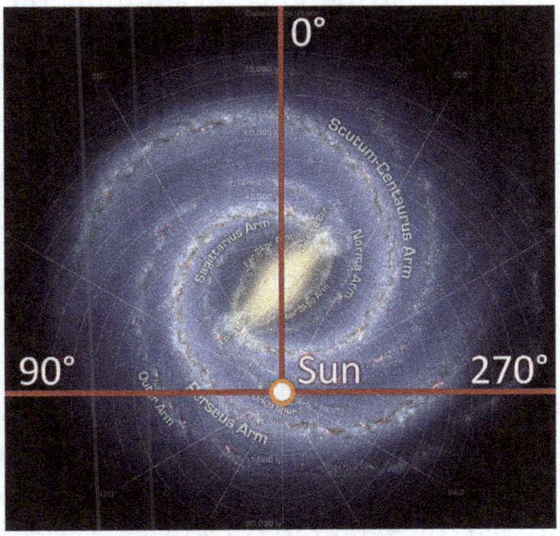

Diagram of the Sun's location in the Milky Way, the angles represent longitudes in the galactic coordinate system.

There are about 208 stars brighter than absolute magnitude 8.5 within a sphere with a radius of 15 parsecs (49 ly) from the Sun, giving a density of one star per 69 cubic parsecs, or one star per 2,360 cubic light-years (from List of nearest bright stars). On the other hand, there are 64 known stars (of any magnitude, not counting 4 brown dwarfs) within 5 parsecs (16 ly) of the Sun, giving a density of about one star per 8.2 cubic parsecs, or one per 284 cubic light-years (from List of nearest stars). This illustrates the fact that there are far more faint stars than bright stars: in the entire sky, there are about 500 stars brighter than apparent magnitude 4 but 15.5 million stars brighter than apparent magnitude 14.

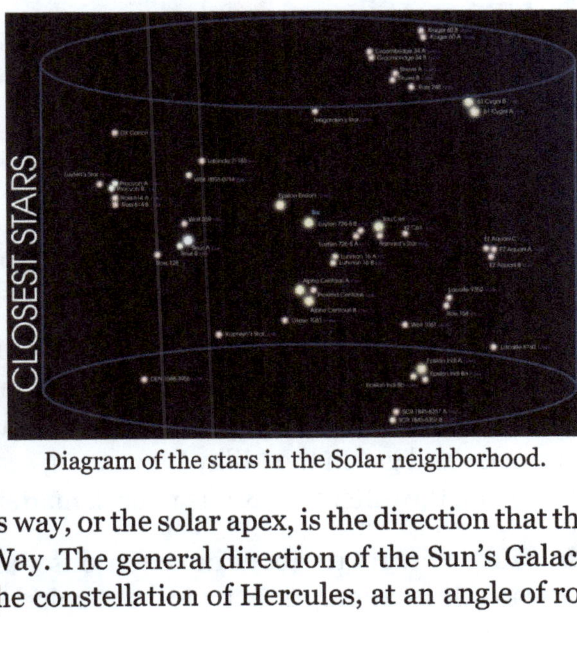

Diagram of the stars in the Solar neighborhood.

The apex of the Sun's way, or the solar apex, is the direction that the Sun travels through space in the Milky Way. The general direction of the Sun's Galactic motion is towards the star Vega near the constellation of Hercules, at an angle of roughly 60 sky degrees to the direction of the Galactic Center. The Sun's orbit about the Milky Way is expected to be roughly elliptical with the addition of perturbations due to the Galactic spiral arms and non-uniform mass distributions. In addition, the Sun passes through the Galactic plane approximately 2.7 times per orbit. This is very similar to how a simple harmonic oscillator works with no drag force (damping) term. These oscillations were until recently thought to coincide with mass lifeform extinction periods on Earth. However, a reanalysis of the effects of the Sun's transit through the spiral structure based on CO data has failed to find a correlation.

It takes the Solar System about 240 million years to complete one orbit of the Milky Way (a galactic year), so the Sun is thought to have completed 18–20 orbits during its lifetime and 1/1250 of a revolution since the origin of humans. The orbital speed of the Solar System about the center of the Milky Way is approximately 220 km/s (490,000 mph) or 0.073% of the speed of light. The Sun moves through the heliosphere at 84,000 km/h (52,000 mph). At this speed, it takes around 1,400 years for the Solar System to travel a distance of 1 light-year, or 8 days to travel 1 AU (astronomical unit). The Solar System is headed in the direction of the zodiacal constellation Scorpius, which follows the ecliptic.

Galactic Rotation

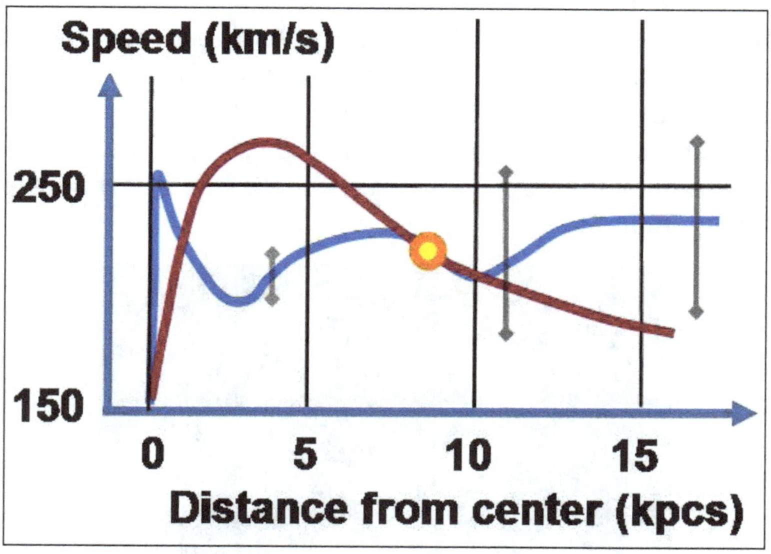

Galaxy rotation curve for the Milky Way – vertical axis is speed of rotation about the galactic center; horizontal axis is distance from the galactic center in kpcs; the sun is marked with a yellow ball; the observed curve of speed of rotation is blue; the predicted curve based upon stellar mass and gas in the Milky Way is red; scatter in observations roughly indicated by gray bars, the difference is due to dark matter.

The stars and gas in the Milky Way rotate about its center differentially, meaning that the rotation period varies with location. As is typical for spiral galaxies, the orbital speed of most stars in the Milky Way does not depend strongly on their distance from the center. Away from the central bulge or outer rim, the typical stellar orbital speed is between 210 ± 10 km/s (470,000 ± 22,000 mph). Hence the orbital period of the typical star is directly proportional only to the length of the path traveled. This is unlike the situation within the Solar System, where two-body gravitational dynamics dominate, and different orbits have significantly different velocities associated with them. The rotation curve (shown in the figure) describes this rotation. Toward the center of the Milky Way the orbit speeds are too low, whereas beyond 7 kpcs the speeds are too high to match what would be expected from the universal law of gravitation.

If the Milky Way contained only the mass observed in stars, gas, and other baryonic (ordinary) matter, the rotation speed would decrease with distance from the center. However, the observed curve is relatively flat, indicating that there is additional mass that cannot be detected directly with electromagnetic radiation. This inconsistency is attributed to dark matter. The rotation curve of the Milky Way agrees with the universal rotation curve of spiral galaxies, the best evidence for the existence of dark matter in galaxies. Alternatively, a minority of astronomers propose that a modification of the law of gravity may explain the observed rotation curve.

Formation

The Milky Way began as one or several small overdensities in the mass distribution in the Universe shortly after the Big Bang. Some of these overdensities were the seeds of globular clusters in which the oldest remaining stars in what is now the Milky Way formed. Nearly half the matter in the Milky Way may have come from other distant galaxies. Nonetheless, these stars and clusters now comprise the stellar halo of the Milky Way. Within a few billion years of the birth of the first stars, the mass of the Milky Way was large enough so that it was spinning relatively quickly. Due to conservation of angular momentum, this led the gaseous interstellar medium to collapse from a roughly spheroidal shape to a disk. Therefore, later generations of stars formed in this spiral disk. Most younger stars, including the Sun, are observed to be in the disk.

Since the first stars began to form, the Milky Way has grown through both galaxy mergers (particularly early in the Milky Way's growth) and accretion of gas directly from the Galactic halo. The Milky Way is currently accreting material from several small galaxies, including two of its largest satellite galaxies, the Large and Small Magellanic Clouds, through the Magellanic Stream. Direct accretion of gas is observed in high-velocity clouds like the Smith Cloud. However, properties of the Milky Way such as stellar mass, angular momentum, and metallicity in its outermost regions suggest it has undergone no mergers with large galaxies in the last 10 billion years. This lack of recent major mergers is unusual among similar spiral galaxies; its neighbour the Andromeda Galaxy appears to have a more typical history shaped by more recent mergers with relatively large galaxies.

According to recent studies, the Milky Way as well as the Andromeda Galaxy lie in what in the galaxy color–magnitude diagram is known as the "green valley", a region populated by galaxies in transition from the "blue cloud" (galaxies actively forming new stars) to the "red sequence" (galaxies that lack star formation). Star-formation activity in green valley galaxies is slowing as they run out of star-forming gas in the interstellar medium. In simulated galaxies with similar properties, star formation will typically have been extinguished within about five billion years from now, even accounting for the expected, short-term increase in the rate of star formation due to the collision between both the Milky Way and the Andromeda Galaxy. In fact, measurements of other galaxies similar to the Milky Way suggest it is among the reddest and brightest spiral galaxies that are still forming new stars and it is just slightly bluer than the bluest red sequence galaxies. In 2019, Scientists at the Aryabhatta Research Institute of Observational Sciences (ARIES) discovered 28 new stars in the Milky Way.

Illustration of a night sky from a hypothetical planet within the Milky Way 10 billion years ago.

Globular clusters are among the oldest objects in the Milky Way, which thus set a lower limit on the age of the Milky Way. The ages of individual stars in the Milky Way can be estimated by measuring the abundance of long-lived radioactive elements such as thorium-232 and uranium-238, then comparing the results to estimates of their original abundance, a technique called nucleocosmochronology. These yield values of about 12.5 ± 3 billion years for CS 31082-001 and 13.8 ± 4 billion years for BD +17° 3248. Once a white dwarf is formed, it begins to undergo radiative cooling and the surface temperature steadily drops. By measuring the temperatures of the coolest of these white dwarfs and comparing them to their expected initial temperature, an age estimate can be made. With this technique, the age of the globular cluster M4 was estimated as 12.7 ± 0.7 billion years. Age estimates of the oldest of these clusters gives a best fit estimate of 12.6 billion years, and a 95% confidence upper limit of 16 billion years.

In November 2018, astronomers reported the discovery of one of the oldest stars in the universe. About 13.5 billion-years-old, 2MASS J18082002-5104378 B is a tiny ultra metal-poor (UMP) star made almost entirely of materials released from the Big Bang, and is possibly one of the very first stars. The discovery of the star in the Milky Way galaxy suggests that the galaxy may be at least 3 billion years older than previously thought.

Several individual stars have been found in the Milky Way's halo with measured ages very close to the 13.80-billion-year age of the Universe. In 2007, a star in the galactic halo, HE 1523-0901, was estimated to be about 13.2 billion years old. As the oldest known object in the Milky Way at that time, this measurement placed a lower limit on the age of the Milky Way. This estimate was made using the UV-Visual Echelle Spectrograph of the Very Large Telescope to measure the relative strengths of spectral lines caused by the presence of thorium and other elements created by the R-process. The line strengths yield abundances of different elemental isotopes, from which an estimate of the age of the star can be derived using nucleocosmochronology. Another star, HD 140283, is 14.5 ± 0.7 billion years old.

According to observations utilizing adaptive optics to correct for Earth's atmospheric distortion, stars in the galaxy's bulge date to about 12.8 billion years old.

The age of stars in the galactic thin disk has also been estimated using nucleocosmochronology. Measurements of thin disk stars yield an estimate that the thin disk formed 8.8 ± 1.7 billion years ago. These measurements suggest there was a hiatus of almost 5 billion years between the formation of the galactic halo and the thin disk. Recent analysis of the chemical signatures of thousands of stars suggests that stellar formation might have dropped by an order of magnitude at the time of disk formation, 10 to 8 billion years ago, when interstellar gas was too hot to form new stars at the same rate as before.

The satellite galaxies surrounding the Milky way are not randomly distributed but seemed to be the result of a break-up of some larger system producing a ring structure 500,000 light-years in diameter and 50,000 light-years wide. Close encounters between galaxies, like that expected in 4 billion years with the Andromeda Galaxy rips off huge tails of gas, which, over time can coalesce to form dwarf galaxies in a ring at an arbitrary angle to the main disc.

Environment

The Milky Way and the Andromeda Galaxy are a binary system of giant spiral galaxies belonging to a group of 50 closely bound galaxies known as the Local Group, surrounded by a Local Void, itself being part of the Virgo Supercluster. Surrounding the Virgo Supercluster are a number of voids, devoid of many galaxies, the Microscopium Void to the "north", the Sculptor Void to the "left", the Bootes Void to the "right" and the Canes-Major Void to the South. These voids change shape over time, creating filamentous structures of galaxies. The Virgo Supercluster, for instance, is being drawn towards the Great Attractor, which in turn forms part of a greater structure, called Laniakea.

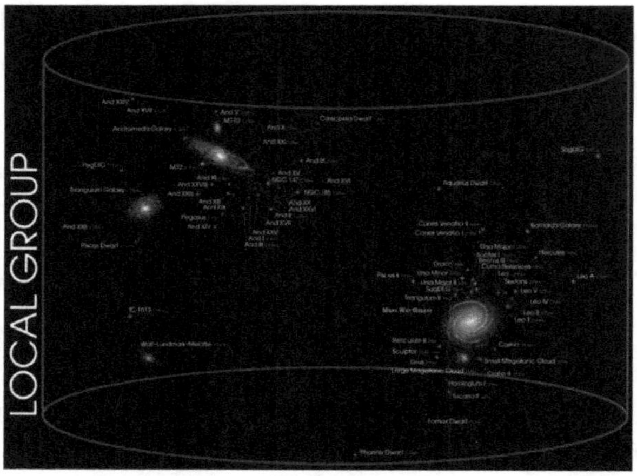

Diagram of the galaxies in the Local Group relative to the Milky Way.

Two smaller galaxies and a number of dwarf galaxies in the Local Group orbit the Milky Way. The largest of these is the Large Magellanic Cloud with a diameter of 14,000 light-years. It has a close companion, the Small Magellanic Cloud. The Magellanic Stream is a stream of neutral hydrogen gas extending from these two small galaxies across 100° of the sky. The stream is thought to have been dragged from the Magellanic Clouds in tidal interactions with the Milky Way. Some of the dwarf galaxies orbiting the Milky Way are Canis Major Dwarf (the closest), Sagittarius Dwarf Elliptical Galaxy, Ursa Minor Dwarf, Sculptor Dwarf, Sextans Dwarf, Fornax Dwarf, and Leo I Dwarf. The smallest dwarf galaxies of the Milky Way are only 500 light-years in diameter. These include Carina Dwarf, Draco Dwarf, and Leo II Dwarf. There may still be undetected dwarf galaxies that are dynamically bound to the Milky Way, which is supported by the detection of nine new satellites of the Milky Way in a relatively small patch of the night sky in 2015. There are also some dwarf galaxies that have already been absorbed by the Milky Way, such as the progenitor of Omega Centauri.

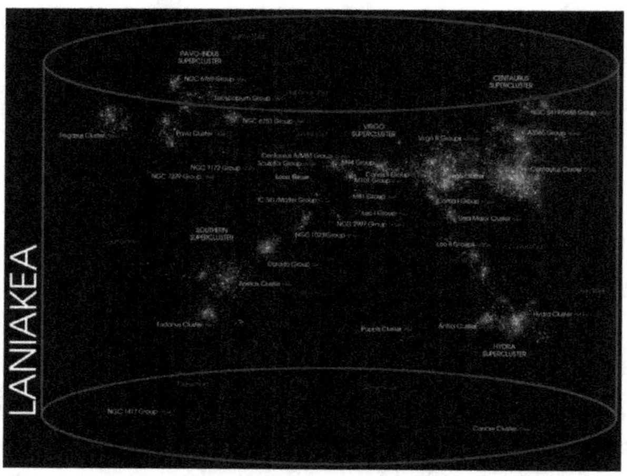

The position of the Local Group within the Laniakea Supercluster.

In 2014 researchers reported that most satellite galaxies of the Milky Way lie in a very large disk and orbit in the same direction. This came as a surprise: according to standard cosmology, the satellite galaxies should form in dark matter halos, and they should be widely distributed and moving in random directions. This discrepancy is still not fully explained.

In January 2006, researchers reported that the heretofore unexplained warp in the disk of the Milky Way has now been mapped and found to be a ripple or vibration set up by the Large and Small Magellanic Clouds as they orbit the Milky Way, causing vibrations when they pass through its edges. Previously, these two galaxies, at around 2% of the mass of the Milky Way, were considered too small to influence the Milky Way. However, in a computer model, the movement of these two galaxies creates a dark matter wake that amplifies their influence on the larger Milky Way.

Current measurements suggest the Andromeda Galaxy is approaching us at 100 to 140 km/s (220,000 to 310,000 mph). In 3 to 4 billion years, there may be an Andromeda–Milky Way collision, depending on the importance of unknown lateral components to the galaxies' relative motion. If they collide, the chance of individual stars colliding with each other is extremely low, but instead the two galaxies will merge to form a single elliptical galaxy or perhaps a large disk galaxy over the course of about a billion years.

Velocity

Although special relativity states that there is no "preferred" inertial frame of reference in space with which to compare the Milky Way, the Milky Way does have a velocity with respect to cosmological frames of reference.

One such frame of reference is the Hubble flow, the apparent motions of galaxy clusters due to the expansion of space. Individual galaxies, including the Milky Way, have peculiar velocities relative to the average flow. Thus, to compare the Milky Way to the Hubble flow, one must consider a volume large enough so that the expansion of the Universe dominates over local, random motions. A large enough volume means that the mean motion of galaxies within this volume is equal to the Hubble flow. Astronomers believe the Milky Way is moving at approximately 630 km/s (1,400,000 mph) with respect to this local co-moving frame of reference. The Milky Way is moving in the general direction of the Great Attractor and other galaxy clusters, including the Shapley supercluster, behind it. The Local Group (a cluster of gravitationally bound galaxies containing, among others, the Milky Way and the Andromeda Galaxy) is part of a supercluster called the Local Supercluster, centered near the Virgo Cluster: although they are moving away from each other at 967 km/s (2,160,000 mph) as part of the Hubble flow, this velocity is less than would be expected given the 16.8 million pc distance due to the gravitational attraction between the Local Group and the Virgo Cluster.

Another reference frame is provided by the cosmic microwave background (CMB). The Milky Way is moving at 552 ± 6 km/s (1,235,000 ± 13,000 mph) with respect to the photons of the CMB, toward 10.5 right ascension, −24° declination (J2000 epoch, near the center of Hydra). This motion is observed by satellites such as the Cosmic Background Explorer (COBE) and the Wilkinson Microwave Anisotropy Probe (WMAP) as a dipole contribution to the CMB, as photons in equilibrium in the CMB frame get blue-shifted in the direction of the motion and red-shifted in the opposite direction.

Andromeda Galaxy

The Andromeda Galaxy, also known as Messier 31, M31, or NGC 224 and originally the Andromeda Nebula, is a spiral galaxy approximately 2.5 million light-years (770 kiloparsecs) from Earth, and the nearest major galaxy to the Milky Way. The galaxy's name stems from the area of the Earth's sky in which it appears, the constellation of Andromeda.

The virial mass of the Andromeda Galaxy is of the same order of magnitude as that of the Milky Way, at 1 trillion solar masses (2.0×10^{42} kilograms). The mass of either galaxy is difficult to estimate with any accuracy, but it was long thought that the Andromeda Galaxy is more massive than the Milky Way by a margin of some 25% to 50%. This has been called into question by a 2018 study which cited a lower estimate on the mass of the Andromeda Galaxy, combined with preliminary reports on a 2019 study estimating a higher mass of the Milky Way. The Andromeda Galaxy has a diameter of about 220,000 ly (67 kpc), making it the largest member of the Local Group at least in terms of extension, if not mass.

The number of stars contained in the Andromeda Galaxy is estimated at one trillion (1×10^{12}), or roughly twice the number estimated for the Milky Way.

The Milky Way and Andromeda galaxies are expected to collide in around 4.5 billion years, merging to form a giant elliptical galaxy or a large lenticular galaxy. With an apparent magnitude of 3.4, the Andromeda Galaxy is among the brightest of the Messier objects making it visible to the naked eye from Earth on moonless nights, even when viewed from areas with moderate light pollution.

Around the year 964, the Persian astronomer Abd al-Rahman al-Sufi was the first to describe the Andromeda Galaxy. He referred to it in his *Book of Fixed Stars* as a "nebulous smear".

Star charts of that period labeled it as the *Little Cloud*. In 1612, the German astronomer Simon Marius gave an early description of the Andromeda Galaxy based on telescopic observations. Pierre Louis Maupertuis conjectured in 1745 that the blurry spot was an island universe. In 1764, Charles Messier cataloged Andromeda as object M31 and incorrectly credited Marius as the discoverer despite it being visible to the naked eye. In 1785, the astronomer William Herschel noted a faint reddish hue in the core region of Andromeda. He believed Andromeda to be the nearest of all the "great nebulae", and based on the color and magnitude of the nebula, he incorrectly guessed that it was no more than 2,000 times the distance of Sirius, or roughly 18,000 ly (5.5 kpc). In 1850, William Parsons, 3rd Earl of Rosse made the first drawing of Andromeda's spiral structure.

In 1864, William Huggins noted that the spectrum of Andromeda differed from that of a gaseous nebula. The spectra of Andromeda displays a continuum of frequencies, superimposed with dark absorption lines that help identify the chemical composition of an object. Andromeda's spectrum is very similar to the spectra of individual stars, and from this, it was deduced that Andromeda has a stellar nature. In 1885, a supernova (known as S Andromedae) was seen in Andromeda, the first and so far only one observed in that galaxy. At the time Andromeda was considered to be a nearby object, so the cause was thought to be a much less luminous and unrelated event called a nova, and was named accordingly; "Nova 1885".

In 1887, Isaac Roberts took the first photographs of Andromeda, which was still commonly thought to be a nebula within our galaxy. Roberts mistook Andromeda and similar spiral nebulae as solar systems being formed. In 1912, Vesto Slipher used spectroscopy to measure the radial velocity of Andromeda with respect to our Solar System—the largest velocity yet measured, at 300 km/s (190 mi/s).

Island Universe

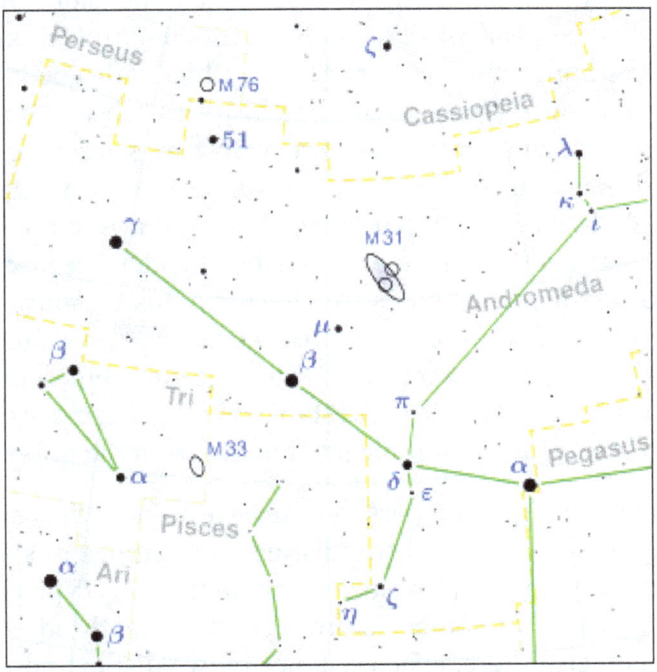

Location of the Andromeda Galaxy (M31) in the Andromeda constellation.

In 1917, Heber Curtis observed a nova within Andromeda. Searching the photographic record, 11 more novae were discovered. Curtis noticed that these novae were, on average, 10 magnitudes fainter than those that occurred elsewhere in the sky. As a result, he was able to come up with a distance estimate of 500,000 ly (3.2×10^{10} AU). He became a proponent of the so-called "island universes" hypothesis, which held that spiral nebulae were actually independent galaxies.

Andromeda Galaxy above the Very Large Telescope. The Triangulum Galaxy is visible on the top.

In 1920, the Great Debate between Harlow Shapley and Curtis took place concerning the nature of the Milky Way, spiral nebulae, and the dimensions of the Universe. To support his claim of the Great Andromeda Nebula being, in fact, an external galaxy, Curtis also noted the appearance of dark lanes within Andromeda which resembled the dust clouds in our own galaxy, as well as historical observations of Andromeda Galaxy's significant Doppler shift. In 1922 Ernst Öpik presented a method to estimate the distance of Andromeda using the measured velocities of its stars. His result placed the Andromeda Nebula far outside our galaxy at a distance of about 450 kpc (1,500 kly). Edwin Hubble settled the debate in 1925 when he identified extragalactic Cepheid variable stars for the first time on astronomical photos of Andromeda. These were made using the 2.5-metre (8 ft 2 in) Hooker telescope, and they enabled the distance of Great Andromeda Nebula to be determined. His measurement demonstrated conclusively that this feature was not a cluster of stars and gas within our own galaxy, but an entirely separate galaxy located a significant distance from the Milky Way.

In 1943, Walter Baade was the first person to resolve stars in the central region of the Andromeda Galaxy. Baade identified two distinct populations of stars based on their metallicity, naming the young, high-velocity stars in the disk Type I and the older, red stars in the bulge Type II. This nomenclature was subsequently adopted for stars within the Milky Way, and elsewhere. (The existence of two distinct populations had been noted earlier by Jan Oort.) Baade also discovered that there were two types of Cepheid variables, which resulted in a doubling of the distance estimate to Andromeda, as well as the remainder of the Universe.

In 1950, radio emission from the Andromeda Galaxy was detected by Hanbury Brown and Cyril Hazard at Jodrell Bank Observatory. The first radio maps of the galaxy were made in the 1950s by John Baldwin and collaborators at the Cambridge Radio Astronomy Group. The core of the Andromeda Galaxy is called 2C 56 in the 2C radio astronomy catalog. In 2009, the first planet may have been discovered in the Andromeda Galaxy. This was detected using a technique called microlensing, which is caused by the deflection of light by a massive object.

The estimated distance of the Andromeda Galaxy from our own was doubled in 1953 when it was discovered that there is another, dimmer type of Cepheid. In the 1990s, measurements of both standard red giants as well as red clump stars from the *Hipparcos* satellite measurements were used to calibrate the Cepheid distances.

Formation and History

The Andromeda Galaxy as seen by NASA's Wide-field Infrared Survey Explorer.

The Andromeda Galaxy was formed roughly 10 billion years ago from the collision and subsequent merger of smaller protogalaxies.

This violent collision formed most of the galaxy's (metal-rich) galactic halo and extended disk. During this epoch, its rate of star formation would have been very high, to the point of becoming a luminous infrared galaxy for roughly 100 million years. Andromeda and the Triangulum Galaxy had a very close passage 2–4 billion years ago. This event produced high rates of star formation across the Andromeda Galaxy's disk—even some globular clusters—and disturbed M33's outer disk.

Over the past 2 billion years, star formation throughout Andromeda's disk is thought to have decreased to the point of near-inactivity. There have been interactions with satellite galaxies like M32, M110, or others that have already been absorbed by Andromeda Galaxy. These interactions have formed structures like Andromeda's Giant Stellar

Stream. A galactic merger roughly 100 million years ago is believed to be responsible for a counter-rotating disk of gas found in the center of Andromeda as well as the presence there of a relatively young (100 million years old) stellar population.

Distance Estimate

At least four distinct techniques have been used to estimate distances from Earth to the Andromeda Galaxy. In 2003, using the infrared surface brightness fluctuations (I-SBF) and adjusting for the new period-luminosity value and a metallicity correction of −0.2 mag dex^{-1} in (O/H), an estimate of 2.57 ± 0.06 million light-years ($1.625\times10^{11} \pm 3.8\times10^{9}$ astronomical units) was derived. A 2004 Cepheid variable method estimated the distance to be 2.51 ± 0.13 million light-years (770 ± 40 kpc). In 2005, an eclipsing binary star was discovered in the Andromeda Galaxy. The binary is two hot blue stars of types O and B. By studying the eclipses of the stars, astronomers were able to measure their sizes. Knowing the sizes and temperatures of the stars, they were able to measure their absolute magnitude. When the visual and absolute magnitudes are known, the distance to the star can be calculated. The stars lie at a distance of $2.52\times10^{6} \pm 0.14\times10^{6}$ ly ($1.594\times10^{11} \pm 8.9\times10^{9}$ AU) and the whole Andromeda Galaxy at about 2.5×10^{6} ly (1.6×10^{11} AU). This new value is in excellent agreement with the previous, independent Cepheid-based distance value. The TRGB method was also used in 2005 giving a distance of $2.56\times10^{6} \pm 0.08\times10^{6}$ ly ($1.619\times10^{11} \pm 5.1\times10^{9}$ AU). Averaged together, these distance estimates give a value of $2.54\times10^{6} \pm 0.11\times10^{6}$ ly ($1.606\times10^{11} \pm 7.0\times10^{9}$ AU). And, from this, the diameter of Andromeda at the widest point is estimated to be 220 ± 3 kly (67,450 ± 920 pc). Applying trigonometry (angular diameter), this is equivalent to an apparent 4.96° angle in the sky.

Mass Estimates

Until 2018, mass estimates for the Andromeda Galaxy's halo (including dark matter) gave a value of approximately $1.5\times10^{12}\,M_{\odot}$ compared to $8\times10^{11}\,M_{\odot}$ for the Milky Way. These contradicted earlier measurements that seemed to indicate that Andromeda Galaxy and the Milky Way are almost equal in mass. In 2018, the equality of mass was re-established by radio results as approximately $8\times10^{11}\,M_{\odot}$ In 2006, Andromeda Galaxy's spheroid was determined to have a higher stellar density than that of the Milky Way and its galactic stellar disk was estimated at about twice the diameter of that of the Milky Way. The total stellar mass of Andromeda Galaxy is estimated to be between $8\times10^{11}\,M_{\odot}$ and $1.1\times10^{12}\,M_{\odot}$., (i.e., around twice as massive as that of the Milky Way) and $1.5\times10^{12}\,M_{\odot}$, with around 30% of that mass in the central bulge, 56% in the disk, and the remaining 14% in the halo. The radio results (similar mass to Milky Way galaxy) should be taken as likeliest as of 2018, although clearly this matter is still under active investigation by a number of research groups worldwide.

As of 2019 current calculations based on escape velocity and dynamical mass measurements puts the Andromeda Galaxy at $0.8 \times 10^{12} M_{\odot}$, which is only half of the Milky Way's new mass calculations done in 2019, at $1.5 \times 10^{12} M_{\odot}$.

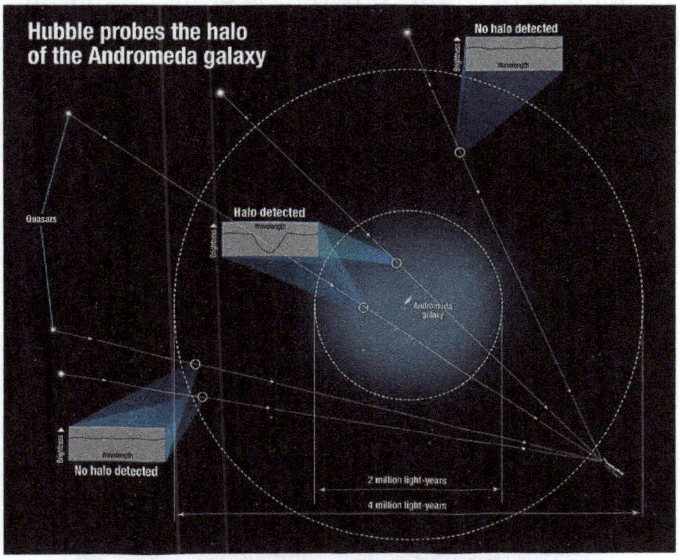

Giant halo around Andromeda Galaxy.

In addition to stars, Andromeda Galaxy's interstellar medium contains at least around $7.2 \times 10^{9} M_{\odot}$ in the form of neutral hydrogen, at least $3.4 \times 10^{8} M_{\odot}$ as molecular hydrogen (within its innermost 10 kiloparsecs), and $5.4 \times 10^{7} M_{\odot}$ of dust.

The Andromeda Galaxy pictured in ultraviolet light.
Illustration showing both the size of each galaxy and the distance between the two galaxies to scale.

Andromeda Galaxy is surrounded by a massive halo of hot gas that is estimated to contain half the mass of the stars in the galaxy. The nearly invisible halo stretches about a million light-years from its host galaxy, halfway to our Milky Way galaxy. Simulations of galaxies indicate the halo formed at the same time as the Andromeda Galaxy. The halo is enriched in elements heavier than hydrogen and helium, formed from supernovae and its properties are those expected for a galaxy that lies in the "green valley" of the Galaxy color–magnitude diagram. Supernovae erupt in Andromeda Galaxy's star-filled disk and eject these heavier elements into space. Over Andromeda Galaxy's lifetime, nearly half of the heavy elements made by its stars have been ejected far beyond the galaxy's 200,000-light-year-diameter stellar disk.

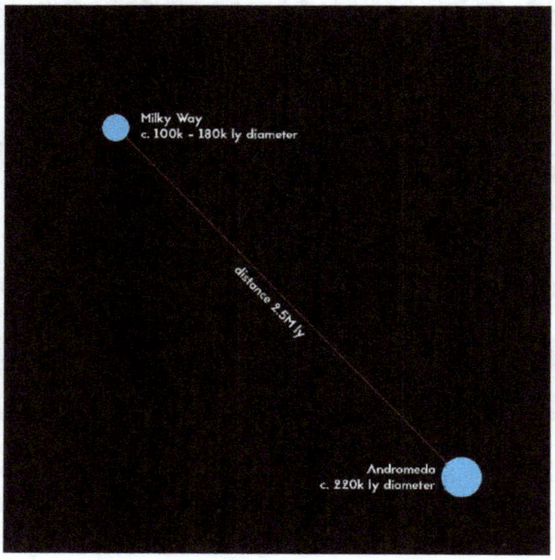

Luminosity Estimates

Compared to the Milky Way, the Andromeda Galaxy appears to have predominantly older stars with ages >7×10^9 years. The estimated luminosity of Andromeda Galaxy, ~2.6×10^{10} $L_\odot$, is about 25% higher than that of our own galaxy. However, the galaxy has a high inclination as seen from Earth and its interstellar dust absorbs an unknown amount of light, so it is difficult to estimate its actual brightness and other authors have given other values for the luminosity of the Andromeda Galaxy (some authors even propose it is the second-brightest galaxy within a radius of 10 mega-parsecs of the Milky Way, after the Sombrero Galaxy, with an absolute magnitude of around -22.21 or close).

An estimation done with the help of Spitzer Space Telescope published in 2010 suggests an absolute magnitude (in the blue) of −20.89 (that with a color index of +0.63 translates to an absolute visual magnitude of −21.52, compared to −20.9 for the Milky Way), and a total luminosity in that wavelength of 3.64×10^{10} $L_\odot$.

The rate of star formation in the Milky Way is much higher, with Andromeda Galaxy producing only about one solar mass per year compared to 3–5 solar masses for the Milky Way. The rate of novae in the Milky Way is also double that of Andromeda Galaxy. This suggests that the latter once experienced a great star formation phase, but is now in a relative state of quiescence, whereas the Milky Way is experiencing more active star formation. Should this continue, the luminosity of the Milky Way may eventually overtake that of Andromeda Galaxy?

According to recent studies, the Andromeda Galaxy lies in what in the galaxy color–magnitude diagram is known as the "green valley", a region populated by galaxies like the Milky Way in transition from the "blue cloud" (galaxies actively forming new stars) to the "red sequence" (galaxies that lack star formation). Star formation activity in green valley galaxies is slowing as they run out of star-forming gas in the interstellar medium. In simulated galaxies with similar properties to Andromeda Galaxy, star formation is expected to extinguish within about five billion years from the now, even accounting for the expected, short-term increase in the rate of star formation due to the collision between Andromeda Galaxy and the Milky Way.

Structure

A Galaxy Evolution Explorer image of the Andromeda Galaxy. The bands of blue-white making up the galaxy's striking rings are neighborhoods that harbor hot, young, massive stars. Dark blue-grey lanes of cooler dust show up starkly against these bright rings, tracing the regions where star formation is currently taking place in dense cloudy cocoons. When observed in visible light, Andromeda Galaxy's rings look more like spiral arms. The ultraviolet view shows that these arms more closely resemble the ring-like structure previously observed in infrared wavelengths with NASA's Spitzer Space Telescope. Astronomers using the latter interpreted these rings as evidence that the galaxy was involved in a direct collision with its neighbor, M32, more than 200 million years ago.

The Andromeda Galaxy seen in infrared by the Spitzer Space Telescope, one of NASA's four Great Space Observatories.

Based on its appearance in visible light, the Andromeda Galaxy is classified as an SA(s) b galaxy in the de Vaucouleurs–Sandage extended classification system of spiral galaxies. However, data from the 2MASS survey showed that Andromeda is actually a barred spiral galaxy, like the Milky Way, with Andromeda's bar oriented along its long axis.

In 2005, astronomers used the Keck telescopes to show that the tenuous sprinkle of stars extending outward from the galaxy is actually part of the main disk itself. This means that the spiral disk of stars in the Andromeda Galaxy is three times larger in diameter than previously estimated. This constitutes evidence that there is a vast, extended stellar disk that makes the galaxy more than 220,000 light-years (67 kiloparsecs) in diameter. Previously, estimates of the Andromeda Galaxy's size ranged from 70,000 to 120,000 light-years (21 to 37 kpc) across.

Image of the Andromeda Galaxy taken by Spitzer in infrared, 24 micrometres.

The galaxy is inclined an estimated 77° relative to Earth (where an angle of 90° would be viewed directly from the side). Analysis of the cross-sectional shape of the galaxy appears to demonstrate a pronounced, S-shaped warp, rather than just a flat disk. A possible cause of such a warp could be gravitational interaction with the satellite galaxies near the Andromeda Galaxy. The Galaxy M33 could be responsible for some warp in Andromeda's arms, though more precise distances and radial velocities are required.

Spectroscopic studies have provided detailed measurements of the rotational velocity of the Andromeda Galaxy as a function of radial distance from the core. The rotational velocity has a maximum value of 225 km/s (140 mi/s) at 1,300 ly (82,000,000 AU) from the core, and it has its minimum possibly as low as 50 km/s (31 mi/s) at 7,000 ly (440,000,000 AU) from the core. Further out, rotational velocity rises out to a radius of 33,000 ly (2.1×10^9 AU), where it reaches a peak of 250 km/s (160 mi/s). The velocities slowly decline beyond that distance, dropping to around 200 km/s (120 mi/s) at 80,000 ly (5.1×10^9 AU). These velocity measurements imply a concentrated mass of about $6\times10^9\ M_\odot$ in the nucleus. The total mass of the galaxy increases linearly out to 45,000 ly (2.8×10^9 AU), then more slowly beyond that radius.

The spiral arms of the Andromeda Galaxy are outlined by a series of HII regions, first studied in great detail by Walter Baade and described by him as resembling "beads on a string". His studies show two spiral arms that appear to be tightly wound, although they are more widely spaced than in our galaxy. His descriptions of the spiral structure, as each arm crosses the major axis of the Andromeda Galaxy, are as follows:

Baade's spiral arms of M31			
Arms (N=cross M31's major axis at north, S=cross M31's major axis at south)	Distance from center (arcminutes) (N*/S*)	Distance from center (kpc) (N*/S*)	Notes
N1/S1	3.4/1.7	0.7/0.4	Dust arms with no OB associations of HII regions.
N2/S2	8.0/10.0	1.7/2.1	Dust arms with some OB associations.
N3/S3	25/30	5.3/6.3	As per N2/S2, but with some HII regions too.
N4/S4	50/47	11/9.9	Large numbers of OB associations, HII regions, and little dust.
N5/S5	70/66	15/14	As per N4/S4 but much fainter.
N6/S6	91/95	19/20	Loose OB associations. No dust visible.
N7/S7	110/116	23/24	As per N6/S6 but fainter and inconspicuous.

Since the Andromeda Galaxy is seen close to edge-on, it is difficult to study its spiral structure. Rectified images of the galaxy seem to show a fairly normal spiral galaxy, exhibiting two continuous trailing arms that are separated from each other by a minimum of about 13,000 ly (820,000,000 AU) and that can be followed outward from a distance of roughly 1,600 ly (100,000,000 AU) from the core. Alternative spiral structures have been proposed such as a single spiral arm or a flocculent pattern of long, filamentary, and thick spiral arms.

The most likely cause of the distortions of the spiral pattern is thought to be interaction with galaxy satellites M32 and M110. This can be seen by the displacement of the neutral hydrogen clouds from the stars.

In 1998, images from the European Space Agency's Infrared Space Observatory demonstrated that the overall form of the Andromeda Galaxy may be transitioning into a ring galaxy. The gas and dust within the galaxy is generally formed into several overlapping rings, with a particularly prominent ring formed at a radius of 32,000 ly (9.8 kpc) from the core, nicknamed by some astronomers the *ring of fire*. This ring is hidden from visible light images of the galaxy because it is composed primarily of cold dust, and most of the star formation that is taking place in the Andromeda Galaxy is concentrated there.

Later studies with the help of the Spitzer Space Telescope showed how Andromeda

Galaxy's spiral structure in the infrared appears to be composed of two spiral arms that emerge from a central bar and continue beyond the large ring. Those arms, however, are not continuous and have a segmented structure.

Close examination of the inner region of the Andromeda Galaxy with the same telescope also showed a smaller dust ring that is believed to have been caused by the interaction with M32 more than 200 million years ago. Simulations show that the smaller galaxy passed through the disk of the Andromeda Galaxy along the latter's polar axis. This collision stripped more than half the mass from the smaller M32 and created the ring structures in Andromeda. It is the co-existence of the long-known large ring-like feature in the gas of Messier 31, together with this newly discovered inner ring-like structure, offset from the barycenter, that suggested a nearly head-on collision with the satellite M32, a milder version of the Cartwheel encounter.

Studies of the extended halo of the Andromeda Galaxy show that it is roughly comparable to that of the Milky Way, with stars in the halo being generally "metal-poor", and increasingly so with greater distance. This evidence indicates that the two galaxies have followed similar evolutionary paths. They are likely to have accreted and assimilated about 100–200 low-mass galaxies during the past 12 billion years. The stars in the extended halos of the Andromeda Galaxy and the Milky Way may extend nearly one third the distance separating the two galaxies.

Nucleus

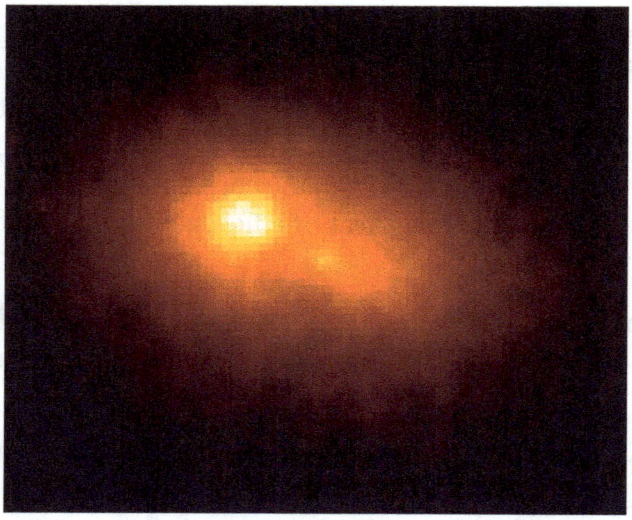

Hubble image of the Andromeda Galaxy core showing possible double structure.

M31 is known to harbor a dense and compact star cluster at its very center. In a large telescope it creates a visual impression of a star embedded in the more diffuse surrounding bulge. In 1991, the Hubble Space Telescope was used to image Andromeda Galaxy's inner nucleus. The nucleus consists of two concentrations separated by 1.5 pc (4.9 ly). The brighter concentration, designated as P1, is offset from the center of the galaxy. The dimmer concentration, P2, falls at the true center of the galaxy and contains a black hole measured at $3–5 \times 10^7 M_\odot$ in 1993, and at $1.1–2.3 \times 10^8 M_\odot$ in 2005. The velocity dispersion of material around it is measured to be ≈ 160 km/s (99 mi/s).

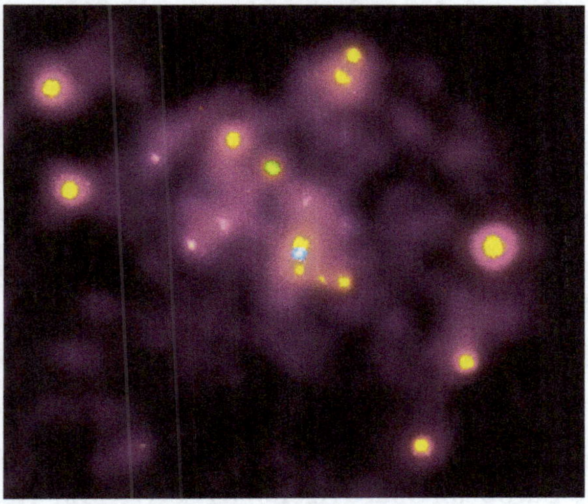

Chandra X-ray telescope image of the center of Andromeda Galaxy. A number of X-ray sources, likely X-ray binary stars, within the galaxy's central region appear as yellowish dots. The blue source at the center is at the position of the supermassive black hole.

It has been proposed that the observed double nucleus could be explained if P1 is the projection of a disk of stars in an eccentric orbit around the central black hole. The eccentricity is such that stars linger at the orbital apocenter, creating a concentration of stars. P2 also contains a compact disk of hot, spectral-class A stars. The A stars are not evident in redder filters, but in blue and ultraviolet light they dominate the nucleus, causing P2 to appear more prominent than P1.

While at the initial time of its discovery it was hypothesized that the brighter portion of the double nucleus is the remnant of a small galaxy "cannibalized" by Andromeda Galaxy, this is no longer considered a viable explanation, largely because such a nucleus would have an exceedingly short lifetime due to tidal disruption by the central black hole. While this could be partially resolved if P1 had its own black hole to stabilize it, the distribution of stars in P1 does not suggest that there is a black hole at its center.

Discrete Sources

Apparently, by late 1968, no X-rays had been detected from the Andromeda Galaxy. A balloon flight on 20 October 1970 set an upper limit for detectable hard X-rays from the Andromeda Galaxy. The Swift BAT all-sky survey successfully detected hard X-rays coming from a region centered 6 arcseconds away from the galaxy center. The emission above 25 keV was later found to be originating from a single source named 3XMM J004232.1+411314, and identified as a binary system where a compact object (a neutron star or a black hole) accretes matter from a star.

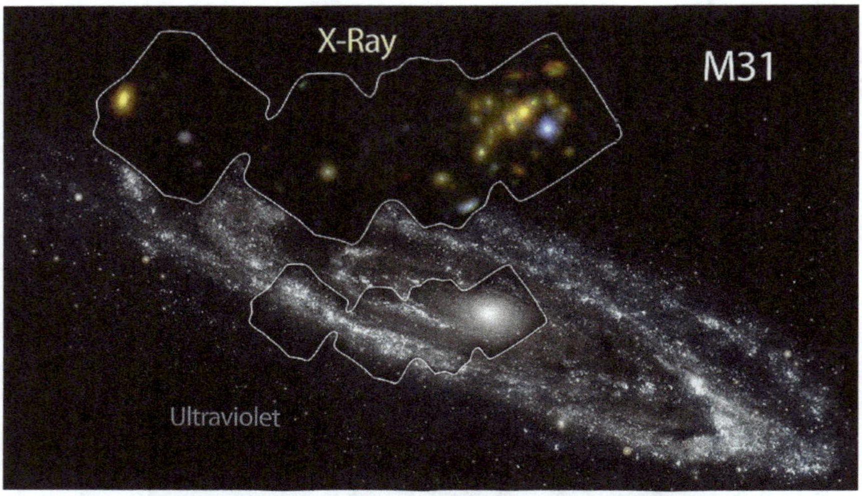

The Andromeda Galaxy in high-energy X-ray and ultraviolet light.

Multiple X-ray sources have since been detected in the Andromeda Galaxy, using observations from the European Space Agency's (ESA) XMM-Newton orbiting observatory. Robin Barnard hypothesized that these are candidate black holes or neutron stars, which are heating the incoming gas to millions of kelvins and emitting X-rays. Neutron stars and black holes can be distinguished mainly by measuring their masses. An observation campaign of NuSTAR space mission identified 40 objects of this kind in the galaxy. In 2012, a microquasar, a radio burst emanating from a smaller black hole was detected in the Andromeda Galaxy. The progenitor black hole is located near the galactic center and has about 10 $M_{\odot}$. It was discovered through data collected by the European Space Agency's XMM-Newton probe and was subsequently observed by NASA's Swift Gamma-Ray Burst Mission and Chandra X-Ray Observatory, the Very Large Array, and the Very Long Baseline Array. The microquasar was the first observed within the Andromeda Galaxy and the first outside of the Milky Way Galaxy.

Artist's concept of the Andromeda Galaxy's core, showing a view across a disk of young, blue stars encircling a supermassive black hole.

Globular Clusters

There are approximately 460 globular clusters associated with the Andromeda Galaxy. The most massive of these clusters, identified as Mayall II, nicknamed Globular One, has a greater luminosity than any other known globular cluster in the Local Group of galaxies. It contains several million stars, and is about twice as luminous as Omega Centauri, the brightest known globular cluster in the Milky Way. Globular One (or G1) has several stellar populations and a structure too massive for an ordinary globular. As a result, some consider G1 to be the remnant core of a dwarf galaxy that was consumed by Andromeda in the distant past. The globular with the greatest apparent brightness is G76 which is located in the south-west arm's eastern half. Another massive globular cluster, named 037-B327 and discovered in 2006 as is heavily reddened by the Andromeda Galaxy's interstellar dust, was thought to be more massive than G1 and the largest cluster of the Local Group; however, other studies have shown it is actually similar in properties to G1.

Star clusters in the Andromeda Galaxy.

Unlike the globular clusters of the Milky Way, which show a relatively low age dispersion, Andromeda Galaxy's globular clusters have a much larger range of ages: from systems as old as the galaxy itself to much younger systems, with ages between a few hundred million years to five billion years.

In 2005, astronomers discovered a completely new type of star cluster in the Andromeda Galaxy. The new-found clusters contain hundreds of thousands of stars, a similar number of stars that can be found in globular clusters. What distinguishes them from the globular clusters is that they are much larger—several hundred light-years across—and hundreds of times less dense. The distances between the stars are, therefore, much greater within the newly discovered extended clusters.

Satellites

Like the Milky Way, the Andromeda Galaxy has satellite galaxies, consisting of over 20 known dwarf galaxies. The best known and most readily observed satellite galaxies are M32 and M110. Based on current evidence, it appears that M32 underwent a close encounter with the Andromeda Galaxy in the past. M32 may once have been a larger galaxy that had its stellar disk removed by M31, and underwent a sharp increase of star formation in the core region, which lasted until the relatively recent past.

Messier 32 is to the left of the center, Messier 110 is to the bottom-right of the center.

M110 also appears to be interacting with the Andromeda Galaxy, and astronomers have found in the halo of the latter a stream of metal-rich stars that appear to have been stripped from these satellite galaxies. M110 does contain a dusty lane, which may indicate recent or ongoing star formation. M32 has a young stellar population as well.

In 2006, it was discovered that nine of the satellite galaxies lie in a plane that intersects the core of the Andromeda Galaxy; they are not randomly arranged as would be expected from independent interactions. This may indicate a common tidal origin for the satellites.

PA-99-N2 Event and Possible Exoplanet in Galaxy

PA-99-N2 was a microlensing event detected in the Andromeda Galaxy in 1999. One of the explanations for this is the gravitational lensing of a red giant with a mass between 0.02 and 3.6 masses of the Sun, which suggested that the star is likely a planet. This possible exoplanet would have mass of 6.34 times that of Jupiters mass. If finally confirmed, it would be the first ever found extragalactic planet. However, anomalies in the event were later found.

Collision with the Milky Way

The Andromeda Galaxy is approaching the Milky Way at about 110 kilometres per second (68 miles per second). It has been measured approaching relative to the Sun at around 300 km/s (190 mi/s) as the Sun orbits around the center of the galaxy at a speed of approximately 225 km/s (140 mi/s). This makes the Andromeda Galaxy one of about

100 observable blueshifted galaxies. Andromeda Galaxy's tangential or sideways velocity with respect to the Milky Way is relatively much smaller than the approaching velocity and therefore it is expected to collide directly with the Milky Way in about 4 billion years. A likely outcome of the collision is that the galaxies will merge to form a giant elliptical galaxy or perhaps even a large disc galaxy. Such events are frequent among the galaxies in galaxy groups. The fate of the Earth and the Solar System in the event of a collision is currently unknown. Before the galaxies merge, there is a small chance that the Solar System could be ejected from the Milky Way or join the Andromeda Galaxy.

Amateur Observing

The Andromeda Galaxy is bright enough to be seen with the naked eye, even with some light pollution. Andromeda is best seen during autumn nights in the Northern Hemisphere, when from mid-latitudes the galaxy reaches zenith (its highest point at midnight) so can be seen almost all night. From the Southern Hemisphere, it is most visible at the same months, that is in spring, and away from our equator does not reach a high altitude over the northern horizon, making it difficult to observe. Binoculars can reveal some larger structures and its two brightest satellite galaxies, M32 and M110. An amateur telescope can reveal Andromeda's disk, some of its brightest globular clusters, dark dust lanes and the large star cloud NGC 206.

Elliptical Galaxy

An elliptical galaxy is a type of galaxy with an approximately ellipsoidal shape and a smooth, nearly featureless image. They are one of the three main classes of galaxy described by Edwin Hubble in his Hubble sequence and 1936 work *The Realm of the Nebulae*, along with spiral and lenticular galaxies. Elliptical (E) galaxies are, together with lenticular galaxies (S0) with their large-scale disks, and ES galaxies with their intermediate scale disks, a subset of the "early-type" galaxy population.

The giant elliptical galaxy ESO 325-G004.

Most elliptical galaxies are composed of older, low-mass stars, with a sparse interstellar medium and minimal star formation activity, and they tend to be surrounded by large numbers of globular clusters. Elliptical galaxies are believed to make up approximately 10%–15% of galaxies in the Virgo Supercluster, and they are not the dominant type of galaxy in the universe overall. They are preferentially found close to the centers of galaxy clusters.

Elliptical galaxies range in size from tens of millions to over one hundred trillion stars. Originally, Edwin Hubble hypothesized that elliptical galaxies evolved into spiral galaxies, which was later discovered to be false, although the accretion of gas and smaller galaxies may build a disk around a pre-existing ellipsoidal structure. Stars found inside of elliptical galaxies are on average much older than stars found in spiral galaxies.

Examples:

- M49.
- M59.
- M60 (NGC 4649).
- M87 (NGC 4486).
- M89.
- M105 (NGC 3379).
- IC 1101, one of the largest galaxies in the observable universe.
- Maffei 1, the closest giant elliptical galaxy.
- CGCG 049-033, known for having the longest galactic jet discovered.

Centaurus A (NGC 5128), a radio galaxy, elliptical/lenticular disputed.

Elliptical galaxies are characterized by several properties that make them distinct from other classes of galaxy. They are spherical or ovoid masses of stars, starved of star-making gases. The smallest known elliptical galaxy is about one-tenth the size of the Milky Way. The motion of stars in elliptical galaxies is predominantly radial, unlike the disks of spiral galaxies, which are dominated by rotation. Furthermore, there is very little interstellar matter (neither gas nor dust), which results in low rates of star formation, few open star clusters, and few young stars; rather elliptical galaxies are dominated by old stellar populations, giving them red colors. Large elliptical galaxies typically have an extensive system of globular clusters.

The dynamical properties of elliptical galaxies and the bulges of disk galaxies are similar, suggesting that they may be formed by the same physical processes, although this remains controversial. The luminosity profiles of both elliptical galaxies and bulges are well fit by Sersic's law, and a range of scaling relations between the elliptical galaxies' structural parameters unify the population.

Elliptical galaxy IC 2006.

Every massive elliptical galaxy contains a supermassive black hole at its center. Observations of 46 elliptical galaxies, 20 classical bulges, and 22 pseudobulges show that each contain a black hole at the center. The mass of the black hole is tightly correlated with the mass of the galaxy, evidenced through correlations such as the M–sigma relation which relates the velocity dispersion of the surrounding stars to the mass of the black hole at the center.

Elliptical galaxies are preferentially found in galaxy clusters and in compact groups of galaxies.

Unlike flat spiral galaxies with organization and structure, elliptical galaxies are more three-dimensional, without much structure, and their stars are in somewhat random orbits around the center.

Sizes and Shapes

Elliptical galaxies vary greatly in both size and mass with diameters ranging from 3000 lightyears to more than 700,000 lightyears, and masses from 10^5 to nearly 10^{13} solar masses. This range is much broader for this galaxy type than for any other. The smallest, the dwarf elliptical galaxies, may be no larger than a typical globular cluster, but contain a considerable amount of dark matter not present in clusters. Most of these small galaxies may not be related to other ellipticals.

The Hubble classification of elliptical galaxies contains an integer that describes how elongated the galaxy image is. The classification is determined by the ratio of the major (a) to the minor (b) axes of the galaxy's isophotes:

$$10\times\left(1-\frac{b}{a}\right)$$

The central galaxy in this image is a gigantic elliptical galaxy designated 4C 73.08.

Thus for a spherical galaxy with *a* equal to *b*, the number is 0, and the Hubble type is E0. While the limit in the literature is about E7, it has been known since 1966 that the E4 to E7 galaxies are misclassified lenticular galaxies with disks inclined at different angles to our line of sight. This has been confirmed through spectral observations revealing the rotation of their stellar disks. Hubble recognized that his shape classification depends both on the intrinsic shape of the galaxy, as well as the angle with which the galaxy is observed. Hence, some galaxies with Hubble type E0 are actually elongated.

The brilliant central object is a supergiant elliptical galaxy, the dominant member of a galaxy cluster with the name MACSJ1423.8+2404. Note the gravitational lensing.

It is sometimes said that there are two physical types of ellipticals: the giant ellipticals with slightly "boxy"-shaped isophotes, whose shapes result from random motion which is greater in some directions than in others (anisotropic random motion); and the "disky" normal and dwarf ellipticals, which contain disks. This is, however, an abuse of the nomenclature, as there two types of early-type galaxy, those with disks and those without. Given the existence of ES galaxies with intermediate-scale disks, it is reasonable to expect that there is continuity from E to ES, and onto the So galaxies with their large-scale stellar disks that dominate the light at large radii.

Dwarf spheroidal galaxies appear to be a distinct class: their properties are more similar to those of irregulars and late spiral-type galaxies.

At the large end of the elliptical spectrum, there is further division, beyond Hubble's classification. Beyond *gE* giant ellipticals, lies D-galaxies and cD-galaxies. These are similar to their smaller brethren, but more diffuse, with large haloes that may as much belong to the galaxy cluster within which they reside than the centrally-located giant galaxy.

Star Formation

The traditional portrait of elliptical galaxies paints them as galaxies where star formation finished after an initial burst at high-redshift, leaving them to shine with only their aging stars. Elliptical galaxies typically appear yellow-red, which is in contrast to the distinct blue tinge of most spiral galaxies. In spirals, this blue color emanates largely from the young, hot stars in their spiral arms. Very little star formation is thought to occur in elliptical galaxies, because of their lack of gas compared to spiral or irregular galaxies. However, in recent years, evidence has shown that a reasonable proportion (~25%) of early-type (E, ES and So) galaxies have residual gas reservoirs and low level star-formation.

Researchers with the Herschel Space Observatory have speculated that the central black holes in elliptical galaxies keep the gas from cooling enough for star formation.

Spiral Galaxy

Spiral galaxies form a class of galaxy originally described by Edwin Hubble in his 1936 work *The Realm of the Nebulae* and, as such, form part of the Hubble sequence. Most spiral galaxies consist of a flat, rotating disk containing stars, gas and dust, and a central concentration of stars known as the bulge. These are often surrounded by a much fainter halo of stars, many of which reside in globular clusters.

Spiral galaxies are named by their spiral structures that extend from the center into the galactic disc. The spiral arms are sites of ongoing star formation and are brighter than the surrounding disc because of the young, hot OB stars that inhabit them.

Roughly two-thirds of all spirals are observed to have an additional component in the form of a bar-like structure, extending from the central bulge, at the ends of which the spiral arms begin. The proportion of barred spirals relative to their barless cousins has likely changed over the history of the Universe, with only about 10% containing bars about 8 billion years ago, to roughly a quarter 2.5 billion years ago, until present, where over two-thirds of the galaxies in the visible universe (Hubble volume) have bars.

The Milky Way is a barred spiral, although the bar itself is difficult to observe from Earth's current position within the galactic disc. The most convincing evidence for the stars forming a bar in the galactic center comes from several recent surveys, including the Spitzer Space Telescope.

Together with irregular galaxies, spiral galaxies make up approximately 60% of galaxies in today's universe. They are mostly found in low-density regions and are rare in the centers of galaxy clusters.

An example of a spiral galaxy, the Pinwheel Galaxy.

The pioneer of studies of the rotation of the Galaxy and the formation of the spiral arms was Bertil Lindblad in 1925. He realized that the idea of stars arranged permanently in a spiral shape was untenable. Since the angular speed of rotation of the galactic disk varies with distance from the centre of the galaxy (via a standard solar system type of gravitational model), a radial arm (like a spoke) would quickly become curved as the galaxy rotates. The arm would, after a few galactic rotations, become increasingly curved and wind around the galaxy ever tighter. This is called the *winding problem*. Measurements in the late 1960s showed that the orbital velocity of stars in spiral galaxies with respect to their distance from the galactic center is indeed higher than expected from Newtonian dynamics but still cannot explain the stability of the spiral structure.

Since the 1970s, there have been two leading hypotheses or models for the spiral structures of galaxies:

- Star formation caused by density waves in the galactic disk of the galaxy.
- The stochastic self-propagating star formation model (SSPSF model) – star formation caused by shock waves in the interstellar medium. The shock waves

are caused by the stellar winds and supernovae from recent previous star formation, leading to self-propagating and self-sustaining star formation. Spiral structure then arises from differential rotation of the galaxy's disk.

These different hypotheses are not mutually exclusive, as they may explain different types of spiral arms.

Density Wave Model

Bertil Lindblad proposed that the arms represent regions of enhanced density (density waves) that rotate more slowly than the galaxy's stars and gas. As gas enters a density wave, it gets squeezed and makes new stars, some of which are short-lived blue stars that light the arms.

Historical Theory of Lin and Shu

The first acceptable theory for the spiral structure was devised by C. C. Lin and Frank Shu in 1964, attempting to explain the large-scale structure of spirals in terms of a small-amplitude wave propagating with fixed angular velocity, that revolves around the galaxy at a speed different from that of the galaxy's gas and stars. They suggested that the spiral arms were manifestations of spiral density waves – they assumed that the stars travel in slightly elliptical orbits, and that the orientations of their orbits is correlated i.e. the ellipses vary in their orientation (one to another) in a smooth way with increasing distance from the galactic center. This is illustrated in the diagram to the right. It is clear that the elliptical orbits come close together in certain areas to give the effect of arms. Stars therefore do not remain forever in the position that we now see them in, but pass through the arms as they travel in their orbits.

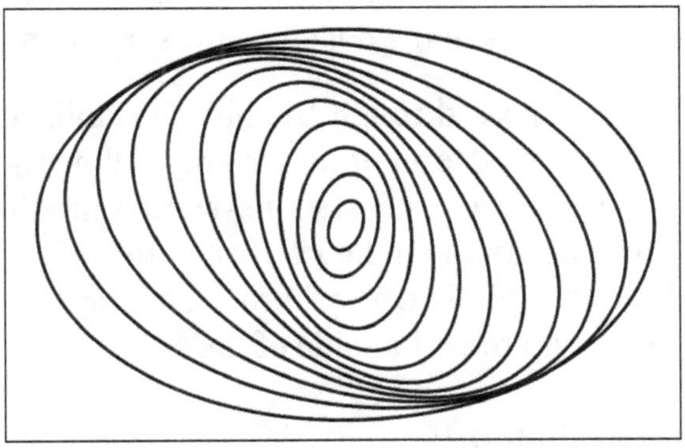

Exaggerated diagram illustrating Lin and Shu's explanation of spiral arms in terms of slightly elliptical orbits.

Star Formation caused by Density Waves

The following hypotheses exist for star formation caused by density waves:

- As gas clouds move into the density wave, the local mass density increases.

Since the criteria for cloud collapse (the Jeans instability) depends on density, a higher density makes it more likely for clouds to collapse and form stars.

- As the compression wave goes through, it triggers star formation on the leading edge of the spiral arms.
- As clouds get swept up by the spiral arms, they collide with one another and drive shock waves through the gas, which in turn causes the gas to collapse and form stars.

The bright galaxy NGC 3810 demonstrates classical spiral structure in this very detailed image from Hubble.

More young Stars in Spiral Arms

Spiral arms appear visually brighter because they contain both young stars and more massive and luminous stars than the rest of the galaxy. As massive stars evolve far more quickly, their demise tends to leave a darker background of fainter stars immediately behind the density waves. This make the density waves much more prominent.

Spiral arms simply appear to pass through the older established stars as they travel in their galactic orbits, so they also do not necessarily follow the arms. As stars move through an arm, the space velocity of each stellar system is modified by the gravitational force of the local higher density. Also the newly created stars do not remain forever fixed in the position within the spiral arms, where the average space velocity returns to normal after the stars depart on the other side of the arm.

Gravitationally Aligned Orbits

Charles Francis and Erik Anderson showed from observations of motions of over 20,000 local stars (within 300 parsecs) that stars do move along spiral arms, and described how mutual gravity between stars causes orbits to align on logarithmic spirals. When the theory is applied to gas, collisions between gas clouds generate the molecular clouds in which new stars form, and evolution towards grand-design bisymmetric spirals is explained.

Structure

Barred spiral galaxy UGC 12158.

Spiral galaxies may consist of several distinct components:

- A flat, rotating disc of stars and interstellar matter of which spiral arms are prominent components.
- A central stellar bulge of mainly older stars, which resembles an elliptical galaxy.
- A bar-shaped distribution of stars.
- A near-spherical halo of stars, including many in globular clusters.
- A supermassive black hole at the very center of the central bulge.
- A near-spherical dark matter halo.

The relative importance, in terms of mass, brightness and size, of the different components varies from galaxy to galaxy.

Spiral Arms

Spiral arms are regions of stars that extend from the center of spiral and barred spiral galaxies. These long, thin regions resemble a spiral and thus give spiral galaxies their name. Naturally, different classifications of spiral galaxies have distinct arm-structures. Sc and SBc galaxies, for instance, have very "loose" arms, whereas Sa and SBa galaxies have tightly wrapped arms (with reference to the Hubble sequence). Either way, spiral arms contain many young, blue stars (due to the high mass density and the high rate of star formation), which make the arms so bright.

NGC 1300 in infrared light.

Bulge

A *bulge* is a large, tightly packed group of stars. The term refers to the central group of stars found in most spiral galaxies, often defined as the excess of stellar light above the inward extrapolation of the outer (exponential) disk light.

Using the Hubble classification, the bulge of Sa galaxies is usually composed of Population II stars, which are old, red stars with low metal content. Further, the bulge of Sa and SBa galaxies tends to be large. In contrast, the bulges of Sc and SBc galaxies are much smaller and are composed of young, blue Population I stars. Some bulges have similar properties to those of elliptical galaxies (scaled down to lower mass and luminosity); others simply appear as higher density centers of disks, with properties similar to disk galaxies.

Many bulges are thought to host a supermassive black hole at their centers. In our own galaxy, for instance, the object called Sagittarius A* is believed to be a supermassive black hole. There are many lines of evidence for the existence of black holes in spiral galaxy centers, including the presence of active nuclei in some spiral galaxies, and dynamical measurements that find large compact central masses in galaxies such as NGC 4258.

Bar

Bar-shaped elongations of stars are observed in roughly two-thirds of all spiral galaxies. Their presence may be either strong or weak. In edge-on spiral (and lenticular) galaxies, the presence of the bar can sometimes be discerned by the out-of-plane X-shaped or (peanut shell)-shaped structures which typically have a maximum visibility at half the length of the in-plane bar.

Spheroid

The bulk of the stars in a spiral galaxy are located either close to a single plane (the galactic plane) in more or less conventional circular orbits around the center of the galaxy (the Galactic Center), or in a spheroidal galactic bulge around the galactic core.

However, some stars inhabit a *spheroidal halo* or *galactic spheroid*, a type of galactic halo. The orbital behaviour of these stars is disputed, but they may exhibit retrograde and highly inclined orbits, or not move in regular orbits at all. Halo stars may be acquired from small galaxies which fall into and merge with the spiral galaxy—for example, the Sagittarius Dwarf Spheroidal Galaxy is in the process of merging with the Milky Way and observations show that some stars in the halo of the Milky Way have been acquired from it.

Spiral galaxy NGC 1345.

Unlike the galactic disc, the halo seems to be free of dust, and in further contrast, stars in the galactic halo are of Population II, much older and with much lower metallicity than their Population I cousins in the galactic disc (but similar to those in the galactic bulge). The galactic halo also contains many globular clusters.

NGC 428 is a barred spiral galaxy, located approximately 48 million light-years away from Earth in the constellation of Cetus.

The motion of halo stars does bring them through the disc on occasion, and a number of small red dwarfs close to the Sun are thought to belong to the galactic halo, for example Kapteyn's Star and Groombridge 1830. Due to their irregular movement around the center of the galaxy, these stars often display unusually high proper motion.

Oldest Spiral Galaxy

The oldest spiral galaxy on file is BX442. At eleven billion years old, it is more than two billion years older than any previous discovery. Researchers think the galaxy's shape is caused by the gravitational influence of a companion dwarf galaxy. Computer models based on that assumption indicate that BX442's spiral structure will last about 100 million years.

Related

In June 2019, citizen scientists through Galaxy Zoo reported that the usual Hubble classification, particularly concerning spiral galaxies, may not be supported, and may need updating.

Distribution of Stars in Spirals

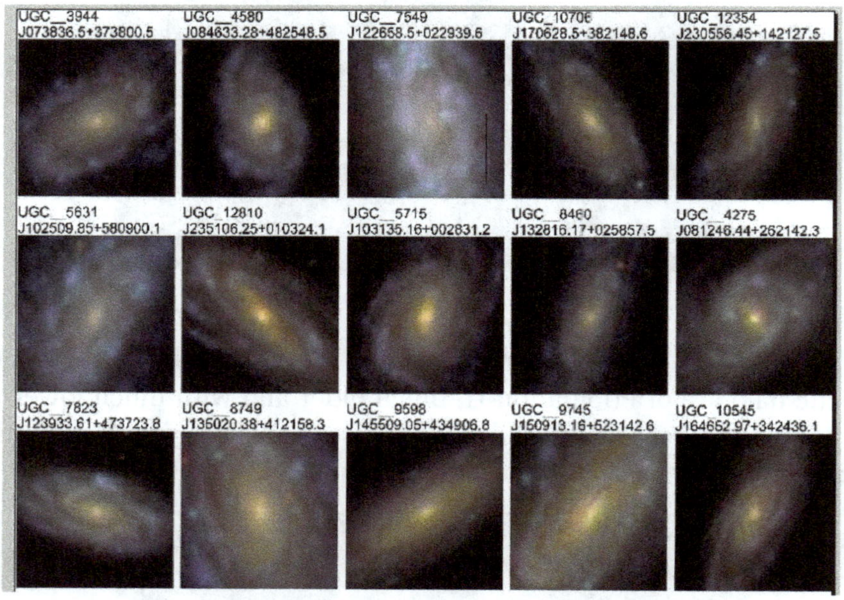

The similar distribution of stars in Spirals.

The stars in spirals are distributed in thin disks radial with intensity profiles such that:

$$I(R) = I_0 e^{-R/h}$$

with h being the disk scale-length; I_0 is the central value; it is useful to define: $R_{opt} = 3.2h$ as the size of the stellar disk, whose luminosity is:

$$L_{tot} = 2\pi I_0 h^2.$$

The spiral galaxies light profiles, in terms of the coordinate R/h, do not depend on galaxy luminosity.

Spiral Nebula

Before it was understood that spiral galaxies existed outside of our Milky Way galaxy, they were often referred to as *spiral nebulae.* The question of whether such objects were separate galaxies independent of the Milky Way, or a type of nebula existing within our own galaxy, was the subject of the Great Debate of 1920, between Heber Curtis of Lick Observatory and Harlow Shapley of Mt. Wilson Observatory. Beginning in 1923, Edwin Hubble observed Cepheid variables in several spiral nebulae, including the so-called "Andromeda Nebula", proving that they are, in fact, entire galaxies outside our own. The term *spiral nebula* has since fallen out of use.

Dwarf Galaxy

A dwarf galaxy is a small galaxy composed of about 100 million up to several billion stars, a small number compared to the Milky Way's 200–400 billion stars. The Large Magellanic Cloud, which closely orbits the Milky Way and contains over 30 billion stars, is sometimes classified as a dwarf galaxy; others consider it a full-fledged galaxy. Dwarf galaxies' formation and activity are thought to be heavily influenced by interactions with larger galaxies. Astronomers identify numerous types of dwarf galaxies, based on their shape and composition.

The Large Magellanic Cloud, a satellite galaxy of the Milky Way.

Current theory states that most galaxies, including dwarf galaxies, form in association with dark matter, or from gas that contains metals. However, NASA's Galaxy Evolution Explorer space probe identified new dwarf galaxies forming out of gases with low metallicity. These galaxies were located in the Leo Ring, a cloud of hydrogen and helium around two massive galaxies in the constellation Leo.

Because of their small size, dwarf galaxies have been observed being pulled toward and ripped by neighbouring spiral galaxies, resulting in galaxy merger.

Dwarf galaxies like NGC 5264 typically possess around a billion stars.

Local Dwarf Galaxies

There are many dwarf galaxies in the Local Group; these small galaxies frequently orbit larger galaxies, such as the Milky Way, the Andromeda Galaxy and the Triangulum Galaxy. A 2007 paper has suggested that many dwarf galaxies were created by galactic tides during the early evolutions of the Milky Way and Andromeda. Tidal dwarf galaxies are produced when galaxies collide and their gravitational masses interact. Streams of galactic material are pulled away from the parent galaxies and the halos of dark matter that surround them. A 2018 study suggests that some local dwarf galaxies formed extremely early, during the Dark Ages within the first billion years after the big bang.

The Phoenix Dwarf Galaxy is a dwarf irregular galaxy, featuring younger stars in its inner regions and older ones at its outskirts.

More than 20 known dwarf galaxies orbit the Milky Way, and recent observations have also led astronomers to believe the largest globular cluster in the Milky Way, Omega Centauri, is in fact the core of a dwarf galaxy with a black hole at its centre, which was at some time absorbed by the Milky Way.

Common Types

UGC 11411 is a galaxy known as an irregular blue compact dwarf (BCD) galaxy.

- Elliptical galaxy: Dwarf elliptical galaxy (dE).
- Dwarf spheroidal galaxy (dSph): Once a subtype of dwarf ellipticals, now regarded as a distinct type.
- Irregular galaxy: Dwarf irregular galaxy (dIrr).
- Spiral galaxy: Dwarf spiral galaxy (dS).
- Magellanic type dwarfs.
- Blue compact dwarf galaxies.
- Ultra-compact dwarf galaxies.

Blue Compact Dwarf Galaxies

In astronomy, a blue compact dwarf galaxy (BCD galaxy) is a small galaxy which contains large clusters of young, hot, massive stars. These stars, the brightest of which are blue, cause the galaxy itself to appear blue in colour. Most BCD galaxies are also classified as dwarf irregular galaxies or as dwarf lenticular galaxies. Because they are composed of star clusters, BCD galaxies lack a uniform shape. They consume gas intensely, which causes their stars to become very violent when forming.

Blue compact dwarf PGC 51017.

BCD galaxies cool in the process of forming new stars. The galaxies' stars are all formed at different time periods, so the galaxies have time to cool and to build up matter to form new stars. As time passes, this star formation changes the shape of the galaxies.

Nearby examples include NGC 1705, NGC 2915, NGC 3353 and UGCA 281.

Ultra-compact Dwarfs

Ultra-compact dwarf galaxies (UCD) are a class of very compact galaxies with very high stellar densities, discovered in the 2000s. They are thought to be on the order of 200 light years across, containing about 100 million stars. It is theorised that these are the cores of nucleated dwarf elliptical galaxies that have been stripped of gas and outlying stars by tidal interactions, travelling through the hearts of rich clusters. UCDs have been found in the Virgo Cluster, Fornax Cluster, Abell 1689, and the Coma Cluster, amongst others. In particular, an unprecedentedly large sample of ~ 100 UCDs has been found in the core region of the Virgo cluster by the Next Generation Virgo Cluster Survey team. The first ever relatively robust studies of the global properties of Virgo UCDs suggest that UCDs have distinct dynamical and structural properties from normal globular clusters. An extreme example of UCD is M60-UCD1, about 54 million light years away, which contains approximately 200 million solar masses within a 160 light year radius; its central region packs in stars about 25 times closer together than stars in Earth's region in the Milky Way. M59-UCD3 is approximately the same size as M60-UCD1 with a half-light radius, r_h, of approximately 20 parsecs but is 40% more luminous with an absolute visual magnitude of approximately −14.6. This makes M59-UCD3 the densest known galaxy. Based on stellar orbital velocities, two UCD in the Virgo Cluster are claimed to have supermassive black holes weighing 13% and 18% of the galaxies' masses.

Starburst Galaxy

A starburst galaxy is a galaxy undergoing an exceptionally high rate of star formation, as compared to the long-term average rate of star formation in the galaxy or the star formation rate observed in most other galaxies. For example, the star formation rate of the Milky Way galaxy is approximately 3 $M_{\odot}$/yr, however, starburst galaxies can experience star formation rates that are more than a factor of 100 times greater. In a starburst galaxy, the rate of star formation is so large that the galaxy will consume all of its gas reservoir, from which the stars are forming, on a timescale much shorter than the age of the galaxy. As such, the starburst nature of a galaxy is a phase, and one that typically occupies a brief period of a galaxy's evolution. The majority of starburst galaxies are in the midst of a merger or close encounter with another galaxy. Starburst galaxies include M82, NGC 4038/NGC 4039 (the Antennae Galaxies), and IC 10.

The Antennae Galaxies are an example of a starburst galaxy occurring from the collision of NGC 4038/NGC 4039.

Starburst galaxies are defined by these three interrelated factors:

- The rate at which the galaxy is currently converting gas into stars (the star-formation rate, or SFR).
- The available quantity of gas from which stars can be formed.
- A comparison of the timescale on which star formation will consume the available gas with the age or rotation period of the galaxy.

Commonly used definitions include:

- Continued star-formation where the current SFR would exhaust the available gas reservoir in much less than the age of the Universe (the Hubble Time).
- Continued star-formation where the current SFR would exhaust the available gas reservoir in much less than the dynamical timescale of the galaxy (perhaps one rotation period in a disk type galaxy).
- The current SFR, normalised by the past-averaged SFR, is much greater than unity. This ratio is referred to as the "birthrate parameter".

Triggering Mechanisms

Mergers and tidal interactions between gas-rich galaxies play a large role in driving starbursts. Galaxies in the midst of a starburst frequently show tidal tails, an indication of a close encounter with another galaxy, or are in the midst of a merger. Interactions between galaxies that do not merge can trigger unstable rotation modes, such as the bar instability, which causes gas to be funneled towards the nucleus and ignites bursts of star formation near the galactic nucleus. It has been shown that there is a strong correlation between the lopsidedness of a galaxy and the youth of its stellar population, with more lopsided galaxies having younger central stellar populations. As lopsidedness can be caused by tidal interactions and mergers between galaxies, this result gives further evidence that mergers and tidal interactions can induce central star formation in a galaxy and drive a starburst.

Types

Artist's impression of a galaxy undergoing a starburst.

Classifying types of starburst galaxies is difficult since starburst galaxies do not represent a specific type in and of themselves. Starbursts can occur in disk galaxies, and irregular galaxies often exhibit knots of starburst spread throughout the irregular galaxy. Nevertheless, astronomers typically classify starburst galaxies based on their most distinct observational characteristics. Some of the categorizations include:

- Blue compact galaxies (BCGs): These galaxies are often low mass, low metallicity, dust-free objects. Because they are dust-free and contain a large number of hot, young stars, they are often blue in optical and ultraviolet colours. It was initially thought that BCGs were genuinely young galaxies in the process of forming their first generation of stars, thus explaining their low metal content. However, old stellar populations have been found in most BCGs, and it is thought that efficient mixing may explain the apparent lack of dust and metals. Most BCGs show signs of recent mergers and/or close interactions. Well-studied BCGs include IZw18 (the most metal poor galaxy known), ESO338-IG04 and Haro11.
    - Blue compact dwarf galaxies (BCD galaxies) are small compact galaxies.
    - Green Pea galaxies (GPs) are small compact galaxies resembling primordial starbursts. They were found by citizen scientists taking part in the Galaxy Zoo project.
- Luminous infrared galaxies (LIRGs):
    - Ultra-luminous Infrared Galaxies ULIRGs): These galaxies are generally extremely dusty objects. The ultraviolet radiation produced by the obscured star-formation is absorbed by the dust and reradiated in the infrared spectrum at wavelengths of around 100 micrometres. This explains the extreme red colours associated with ULIRGs. It is not known for sure that the UV

radiation is produced purely by star-formation, and some astronomers believe ULIRGs to be powered (at least in part) by active galactic nuclei (AGN). X-ray observations of many ULIRGs that penetrate the dust suggest that many starburst galaxies are double-cored systems, lending support to the hypothesis that ULIRGs are powered by star-formation triggered by major mergers. Well-studied ULIRGs include Arp 220.

    - Hyperluminous Infrared galaxies (HLIRGs), sometimes called Submillimeter galaxies.

- Wolf-Rayet galaxies (WR galaxies), galaxies where a large portion of the bright stars are Wolf-Rayet stars. The Wolf-Rayet phase is a relatively short-lived phase in the life of massive stars, typically 10% of the total life-time of these stars and as such any galaxy is likely to contain few of these. However, because the stars are both very luminous and have very distinctive spectral features, it is possible to identify these stars in the spectra of entire galaxies and doing so allows good constraints to be placed on the properties of the starbursts in these galaxies.

References

- Galaxies, science-space-universe: nationalgeographic.com, Retrieved 17 July, 2020
- Rix, Hans-Walter; Bovy, Jo (2013). "The Milky Way's Stellar Disk". The Astronomy and Astrophysics Review. 21: 61. arXiv:1301.3168. Bibcode:2013A&ARv..21...61R. doi:10.1007/s00159-013-0061-8
- "The Need for Speed: Escape velocity and dynamical mass measurements of the Andromeda Galaxy". Monthly Notices of the Royal Astronomical Society. 475 (3): 4043–4054. arXiv:1801.03949. Bibcode:2018MNRAS.475.4043K. doi:10.1093/mnras/sty082. ISSN 0035-8711
- Francis, C.; Anderson, E. (2009). "Galactic spiral structure". Proceedings of the Royal Society A: Mathematical, Physical and Engineering Sciences. 465 (2111): 3425–3446. arXiv:0901.3503. Bibcode:2009RSPSA.465.3425F. doi:10.1098/rspa.2009.0036
- Metz, M.; Kroupa, P. (2007). "Dwarf-spheroidal satellites: are they of tidal origin?". Monthly Notices of the Royal Astronomical Society. 376 (1): 387–392. arXiv:astro-ph/0701289. Bibcode:2007MNRAS.376..387M. doi:10.1111/j.1365-2966.2007.11438.x
- Kennicutt, R. C.; Evans, N. J. (2012). "Star Formation in the Milky Way and Nearby Galaxies". Annual Review of Astronomy and Astrophysics. 50: 531–608. arXiv:1204.3552. Bibcode:2012ARA&A..50..531K. doi:10.1146/annurev-astro-081811-125610
- Quasar, science: britannica.com, Retrieved 25 August, 2020

Permissions

All chapters in this book are published with permission under the Creative Commons Attribution Share Alike License or equivalent. Every chapter published in this book has been scrutinized by our experts. Their significance has been extensively debated. The topics covered herein carry significant information for a comprehensive understanding. They may even be implemented as practical applications or may be referred to as a beginning point for further studies.

We would like to thank the editorial team for lending their expertise to make the book truly unique. They have played a crucial role in the development of this book. Without their invaluable contributions this book wouldn't have been possible. They have made vital efforts to compile up to date information on the varied aspects of this subject to make this book a valuable addition to the collection of many professionals and students.

This book was conceptualized with the vision of imparting up-to-date and integrated information in this field. To ensure the same, a matchless editorial board was set up. Every individual on the board went through rigorous rounds of assessment to prove their worth. After which they invested a large part of their time researching and compiling the most relevant data for our readers.

The editorial board has been involved in producing this book since its inception. They have spent rigorous hours researching and exploring the diverse topics which have resulted in the successful publishing of this book. They have passed on their knowledge of decades through this book. To expedite this challenging task, the publisher supported the team at every step. A small team of assistant editors was also appointed to further simplify the editing procedure and attain best results for the readers.

Apart from the editorial board, the designing team has also invested a significant amount of their time in understanding the subject and creating the most relevant covers. They scrutinized every image to scout for the most suitable representation of the subject and create an appropriate cover for the book.

The publishing team has been an ardent support to the editorial, designing and production team. Their endless efforts to recruit the best for this project, has resulted in the accomplishment of this book. They are a veteran in the field of academics and their pool of knowledge is as vast as their experience in printing. Their expertise and guidance has proved useful at every step. Their uncompromising quality standards have made this book an exceptional effort. Their encouragement from time to time has been an inspiration for everyone.

The publisher and the editorial board hope that this book will prove to be a valuable piece of knowledge for students, practitioners and scholars across the globe.

Index